ADVANCED LEVEL CHEMISTRY

Second edition

Max Baker
Ted Lister
Janet Renshaw

Series editors
Ted Lister and Janet Renshaw

Stanley Thornes (Publishers) Ltd

Contents

Acknowledgements

The authors thank those pupils at Trinity School who trialled the materials.

The authors and publisher would like to acknowledge the contribution of Jan King to the first edition.

The authors would like to thank the following for their kind permission to reproduce photographs: Educational Development Center 10.9, 10.12; Claire Starkey 1.3, 4.1, 6.3; Ken Eward/Science Photo Library 7.5.

First published in 1992 in Great Britain by
Simon and Schuster Education

Second edition published in 1996 by Stanley Thornes (Publishers) Ltd
Delta Place
27 Bath Road
Cheltenham GL53 7TH
United Kingdom

00 / 10 9 8 7 6 5

A catalogue record for this book is available from the British Library

ISBN 0-7487-2334-X

Typeset by Techset Composition Ltd
Printed and bound at Redwood Books, Trowbridge, Wiltshire

Introduction

This book is designed to help you to get ready for a post-16 course in chemistry: Advanced Level, Advanced Supplementary Level, BTEC, GNVQ, Scottish Higher, etc. The precise course you will be following doesn't matter because this book stresses the principles of chemistry which are the same for any course. You could use it before you begin your advanced course or during the first part of that course. You will also find it useful if you are aiming for the higher level GCSE papers. You can also refer to it during your advanced course – especially the chapters on essential physics and maths.* *Access to Advanced Level: Chemistry* has been designed so that you can work through it on your own and the answers to the exercises are at the end of the book. However, cheating won't help your understanding!

How to use this book

Teaching yourself how to do something needs confidence, and often that is the one thing that you don't have. We suggest that you work through each section slowly and don't move on to the next section until you have correctly answered the exercises. If you are getting most of them right, you are doing well. What if you aren't? One resource is a standard chemistry textbook. One of the skills you will have to develop for advanced study is independent learning. There are a variety of approaches to explaining concepts and ours may not always be the best for you. At advanced level (and beyond), referring to textbooks and reading on your own initiative is going to give you a valuable skill and that essential ingredient **confidence** in your ability to learn by yourself.

Good luck and enjoy your chemistry!

Max Baker
Ted Lister
Janet Renshaw

* There is a glossary of important terms on p. 94 for you to refer to.

Chapter 1 The kinetic–particle theory

You should be able to recall some simple experiments which suggest that matter is made of moving particles.
Note 5×10^{-10} is another way of writing 0.000 000 000 5. If you are not familiar with numbers expressed in this way, then refer to Chapter 11.

Moving particles

One of the most basic ideas in science is that all substances are made up of tiny particles (atoms, molecules and ions). These particles are in constant motion and this leads to the name **kinetic–particle theory**, the word kinetic meaning moving. A typical atom is about 5×10^{-10} m across and may be moving at speeds of 1000 km/hour. The three states of matter differ in the spacing, movement and arrangement of the particles as shown in Table 1.1. When a substance is heated, the particles move faster (gain kinetic energy) but do not change in size.

Table 1.1 Particles in solid, liquid and gas

	Solid	Liquid	Gas
Arrangement	regular	random	random
Spacing	touching	close	far apart
Movement	vibrating	rapid jostling	very rapid

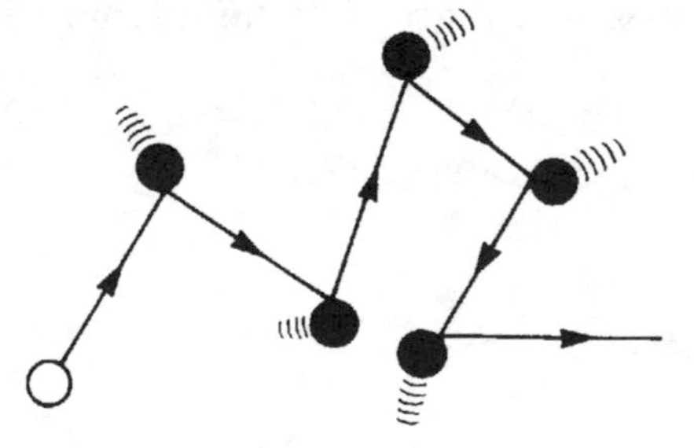

1.1 The path of a gas particle. It will make thousands of collisions every second.

Using the kinetic theory

The kinetic theory enables us to explain the properties of the three states of matter and a number of other things of importance to chemists.

Shape

The particles in a solid must be strongly held together so that solids have a definite shape. In liquids, the particles are slightly further apart and more loosely held so liquids fill the bottom of their containers. Particles in a gas are in rapid, random motion so gases completely fill their containers.

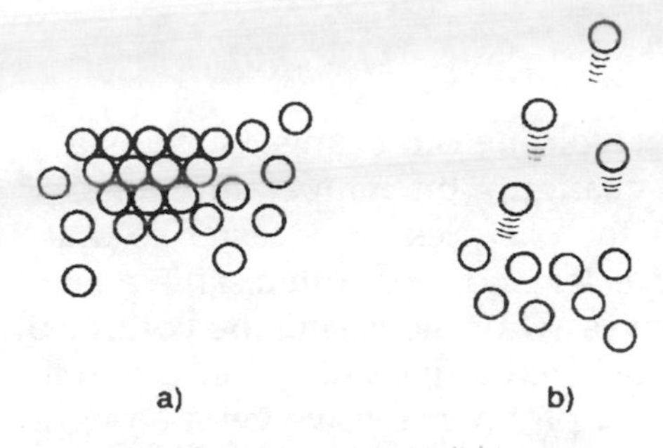

1.2 Pure substances a) melt b) boil at definite temperatures.

Change of state

Particles in solids are vibrating about fixed points. If the solid is heated, the particles gain more energy and vibrate more. Eventually, the particles are vibrating so much that the regular arrangement breaks down and the particles become free to move around. The solid has melted and become liquid. Further heating transfers more energy to the particles until at a particular temperature they become able to move completely freely. The spacing between the particles becomes very large indeed. The liquid has boiled – turned into a gas.

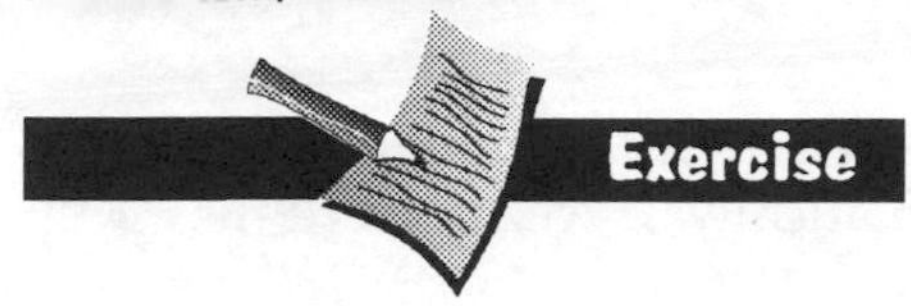

1 Use the kinetic theory and the information in Table 1 to explain why most solids expand when they are heated.

Gases

Gas pressure

All gases exert pressure on the walls of their containers. This is caused by the fast-moving gas particles which bombard the container walls. This pressure is not noticeable in everyday situations, because most containers have gas inside and out. This means that there is an equal and opposite bombardment on both sides and the two cancel out. However, if we use a pump to remove the air from inside an oil can, the effect of the pressure becomes obvious – see Fig 1.3.

1.3 An oil can before and after air has been pumped out of it.

Pressure and temperature

Gas pressure increases with temperature. At higher temperatures, the particles move faster and therefore hit the walls of the container both harder and more often. If we measure the pressure of a fixed mass of gas at different temperatures and plot a graph, we find that it is a straight line – see Fig 1.4.

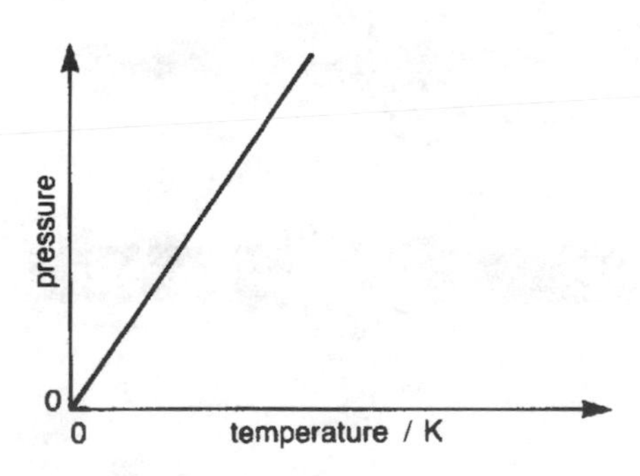

1.4 The straight line shows that pressure is directly proportional to temperature.

Pressure and volume

2 A syringe full of air is connected to a pressure gauge as shown in Fig 1.5

What do you think will happen to the pressure if we squeeze the plunger in so that the gas volume is halved? Try to explain your answer in terms of what the particles of the gas are doing. Hint: how often do they hit the walls, and how hard do they hit the walls after squeezing compared with before?

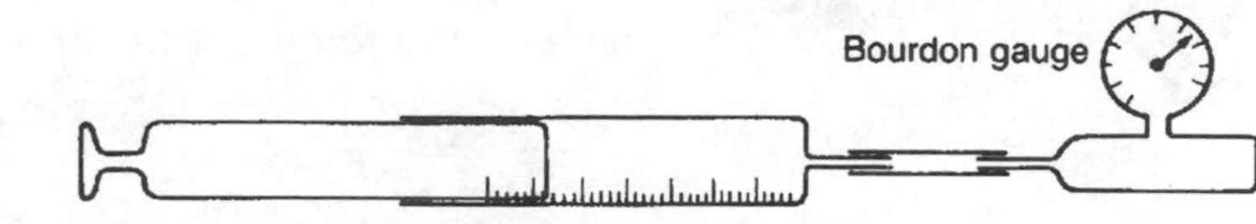

1.5 A syringe of air connected to a pressure gauge.

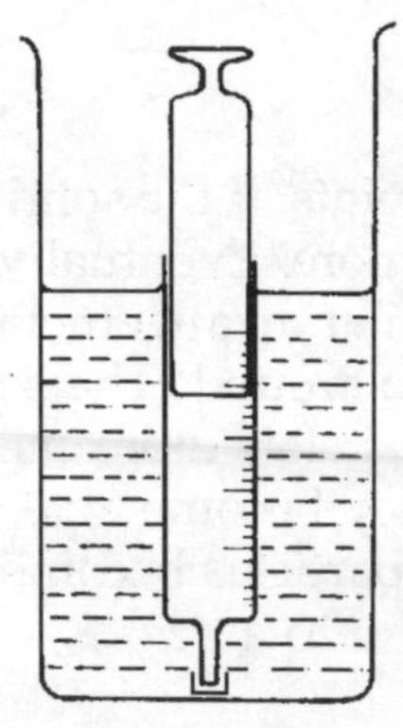

1.6 Changing the temperature of the gas in the syringe will change its volume.

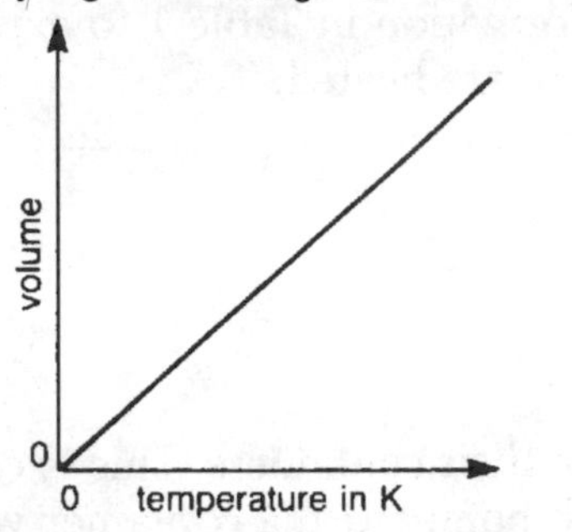

1.7 The straight line shows that the volume of a gas is directly proportional to the temperature. At 0 K the graph predicts that a gas would have no volume!

Volume and temperature

If a gas is kept in a container whose volume can change, such as a syringe (see Fig 1.6) – we find that changing the temperature affects the volume. Heating the gas makes the particles move faster so that they hit the walls of the container harder and more often. This pushes the plunger of the syringe which moves away, increasing the volume of the gas. If we measure the volume of a fixed mass of gas at different temperatures at constant pressure and plot a graph we get a straight line. See Fig 1.7.

The gas laws

The three quantities pressure, P, volume, V, and temperature, T, of a fixed mass of gas are linked mathematically. If we measure the temperature in degrees kelvin, then:

the volume is proportional to the temperature ($V \propto T$) if we keep the pressure constant. This is called **Charles' Law**.

the pressure is proportional to the temperature ($P \propto T$) if we keep the volume constant. This is sometimes called the **Constant Volume Law**.

the product $P \times V$ is always constant if we keep the temperature constant. This is called **Boyle's Law**.

Note The **kelvin scale of temperature** has its zero at $-273\,°C$. We believe that this is the lowest possible temperature. It is often called absolute zero. At this temperature, molecules have no energy at all.

To convert °C to degrees kelvin (written K, not °K) we add 273 and to convert K to °C we subtract 273. Remember that on the Kelvin scale there are no negative temperatures.

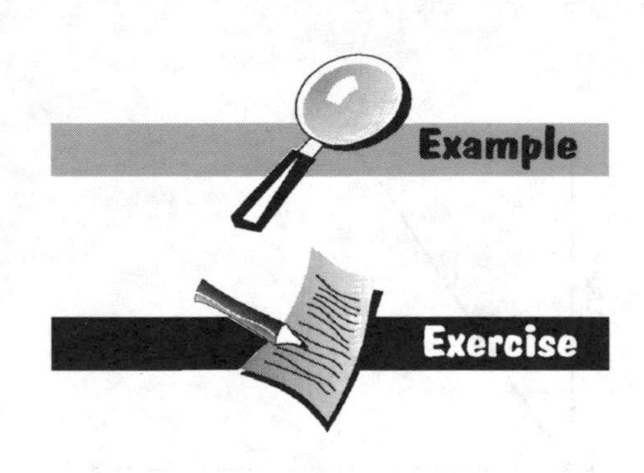

Convert 270 K to °C.

$$270 - 273 = -3\,°C$$

3 a Convert the following temperatures to K: 100 °C, 0 °C, 500 °C.
  b Convert the following temperatures to °C: 500 K, 473 K, 200 K.

The three relationships above can be combined into one equation:

$$\frac{PV}{T} = \text{constant } (T \text{ in kelvin})$$

It follows that

$$\frac{P_1V_1}{T_1} = \frac{P_2V_2}{T_2}$$

where subscripts 1 and 2 refer to particular sets of conditions.

What is the volume of 20 cm^3 of a gas at constant pressure P, if we increase the temperature from 20 °C to 40°C?

$$\frac{P_1V_1}{T_1} = \frac{P_2V_2}{T_2}$$

and $P_1 = P_2 = P$, $V_1 = 20\text{ cm}^3$, $T_1 = 293$ K and $T_2 = 313$ K.

So $$\frac{P \times 20}{293} = \frac{P \times V_2}{313}$$

and rearranging:

$$V_2 = \frac{20 \times 313}{293} \text{ cm}^3$$
$$= 21.4 \text{ cm}^3$$

4 What would the new volume be if we increased the temperature a further 10 °C at the same pressure?

Reaction rates

Catalysts also change reaction rates.

Chemical reactions can only happen if the reacting particles collide with one another. The more often and harder they collide, the faster the reaction will be. This means that the following factors will increase reaction rates:

a increasing the temperature

b increasing the surface area of a solid reactant

c increasing the pressure of a gas

d increasing the concentration of a solution.

5 Explain how each of the above factors increases the rate of a reaction in terms of what the particles are doing.

Diffusion

Diffusion is the word used to describe the fact that gases and liquids tend to mix of their own accord (i.e. without draughts or stirring). For example we can soon tell (by smelling) if the gas has been left on in the kitchen. Diffusion happens because particles are moving fast and there are gaps between the particles in both the solid and the liquid state.

Entropy

An interesting point is that while gases mix completely and readily of their own accord, they will not unmix. We never have to worry about the oxygen and nitrogen in the air unmixing so that all the oxygen goes over to one side of the room and all the nitrogen to the other! This simple observation has a lot to tell us about why some chemical changes happen and others do not. Changes which involve going from order (like two separate gases) to disorder (like a mixture of gases) tend to happen of their own accord. This is rather like the fact that it is easy to mix up a pack of cards but much more effort is needed to arrange them in order. The degree of disorder in a chemical system is called its **entropy**.

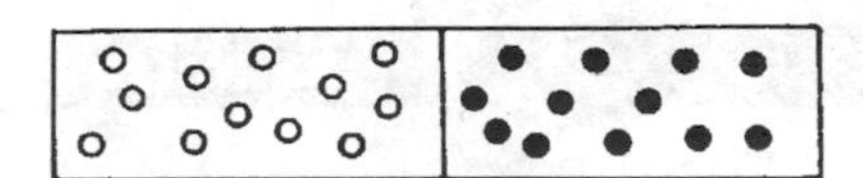

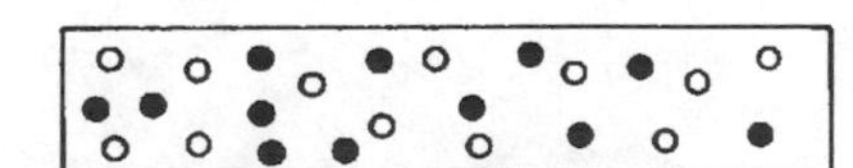

1.8 The separate gases have more order than the mixture, so gases mix of their own accord.

Chapter 2

Atomic structure and bonding

STARTING POINTS

- Do you understand the following terms: **element, compound, atom, relative atomic mass**? Write a sentence or two (no more) to explain what you mean by each term. Then check the glossary at the back of the book.
- Expressing numbers as powers of 10. Do you know that 1×10^{-3} is another way of writing 0.001? If you are not familiar with this, see Chapter 11.

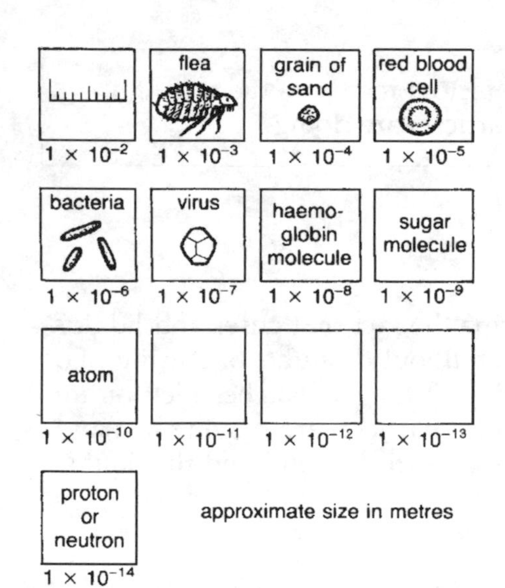

2.1 An idea of the relative size of the atom.

A closer look at atoms

Sub-atomic particles

Scientists believe that atoms are made up of three main sub-atomic particles, the **proton**, the **neutron** and the **electron**. Their properties are summarised in Table 2.1.

The information in the **Periodic Table** enables us to find the number of protons, neutrons, and electrons in any atom. Since the electrons have practically no mass, the mass of the atom is simply the sum of the masses of the protons and neutrons.

Table 2.1 Sub-atomic particles

	proton	**neutron**	**electron**
approximate mass in u*	1	1	1/1840 (approximately 0)
electric charge	+1	0	−1

* 1 u is approximately 1.6×10^{-27} kg

The **atomic number** is the number of protons in an atom. It tells us what element we have because the number of protons is the same as the number of electrons and the electrons are responsible for the chemical reactions of an element. For example, in sodium, Na:

23 ← relative atomic mass, A_r (= protons + neutrons)
Na
11 ← atomic number, Z (= protons = electrons)

So in an atom of sodium, Na:

number of protons = 11
number of electrons = 11
number of neutrons = 23 − 11 = 12

1 Work out the numbers of protons, neutrons and electrons in the following atoms: carbon, C; lithium, Li; uranium, U; manganese, Mn.

2 a What is unique about the make-up of the sub-atomic particles in a hydrogen atom, H?

b What seems strange about the make-up of the sub-atomic particles in a chlorine atom, Cl?

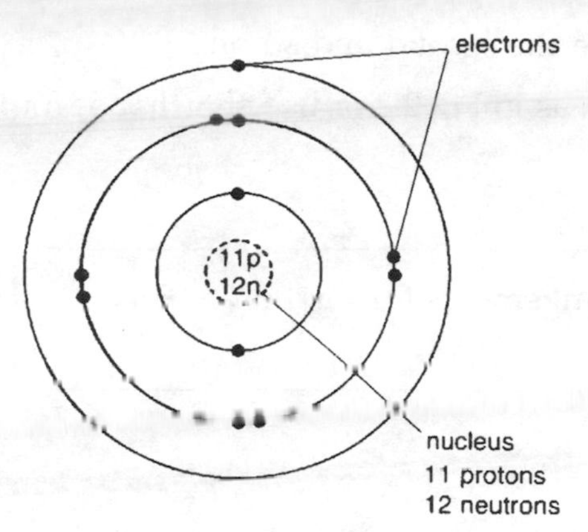

2.2 The sub-atomic particles in a sodium atom.

Arrangement of sub-atomic particles

These sub-atomic particles are arranged as shown in Fig 2.2 to make up the atom. The protons and neutrons are tucked away in the centre of the atom called the **nucleus**, held together by the strong nuclear force. The electrons orbit the nucleus in a series of orbits or shells of increasing distance from the nucleus.

The electrons are held to the nucleus by the attraction of opposite electric charges but they do not fall into it because of their orbital motion.

The scale of the atom

A typical atom has a nucleus which is about 1×10^{-14} m across while the atom itself is about 1×10^{-10} m across. This makes the nucleus some ten thousand times smaller than the whole atom. This would make an atom the size of a soccer pitch have a nucleus the size of a marble.

Isotopes, the strange case of chlorine

Look back at your answer to question 2b. There are not really any $\frac{1}{2}$ neutrons in the nucleus of a chlorine atom. The relative atomic mass shown in the Periodic Table is an average of two sorts of chlorine atom that exist. These are identified by their relative atomic masses as ^{35}Cl and ^{37}Cl. Both types of atom have the same atomic number, 17. They differ in that ^{35}Cl has 18 neutrons while ^{37}Cl has 20. ^{35}Cl atoms are three times as common as ^{37}Cl atoms. This means that the total relative atomic mass of four chlorine atoms picked at random would be

$$(3 \times 35) + (1 \times 37) = 105 + 37 = 142$$

This gives the average relative atomic mass of each atom as $142/4 = 35.5$

Atoms of the same element which have different numbers of neutrons are called **isotopes**.

3 What would be the numbers of protons, neutrons and electrons, in the following isotopes?

a) $^{12}_{6}C$, $^{13}_{6}C$, $^{14}_{6}C$ b) $^{1}_{1}H$, $^{2}_{1}H$, $^{3}_{1}H$

4 Neon, Ne, has isotopes $^{20}_{10}Ne$ and $^{22}_{10}Ne$. They exist in the approximate ratio 9 : 1. What will be the average relative atomic mass of neon?

- In the Periodic Table in this book (and many others) most of the relative atomic masses have been rounded to the nearest whole number.

Arrangement of electrons

The electron shells can hold different numbers of electrons. In order of increasing distance from the nucleus, these are:

First shell	up to 2
Second shell	up to 8
Third shell	up to 8 with reserve space for 10 more

The first shell is always filled before the second and so on.

We can draw electron arrangements as in Fig 2.2 or use shorthand, for example sodium: Na, 2,8,1.

5 Draw or write the electron arrangements for a boron, B; b fluorine, F; c lithium, Li.

Uncovering the structure of the atom

The full story of this would take up too much room for this book but it is a fascinating tale and well worth looking up. Below is a list of some of the main events and dates. The story is, of course, still going on.

1803 Dalton's model of the atom as a small billiard ball (the idea used in kinetic theory).
Late 19th century Discovery of radioactive elements (their break-up reveals that atoms have structure).
1897 Joseph John ('J J') Thomson discovers the electron (a negatively charged constituent of all atoms).
1890s J J Thomson's model of the atom as a positive sphere embedded with rings of electrons (the currant bun model).
1909 Geiger, Marsden and Rutherford's alpha scattering experiment shows that atoms have a nucleus. (see Fig 2.3)
1911 Rutherford's model of the atom has electrons orbiting the nucleus.
1913 Bohr suggests electron shells.
1920 Rutherford names the proton.
1920s Electrons considered as clouds of charge rather than as particles.
1932 Chadwick discovers the neutron.
To the present day Discovery of even more sub-atomic particles such as **muons, quarks, charmed particles** and so on. You might like to look some of these up.

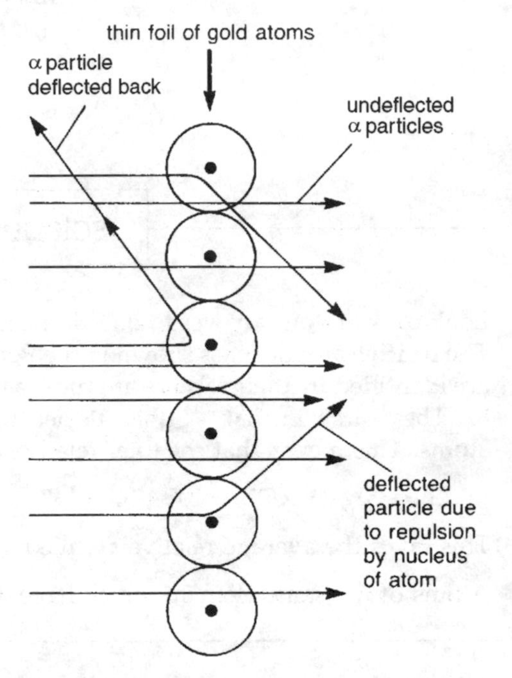

2.3 Most of the α particles passed through the thin foil but a few were deflected back by collision with the nuclei of the gold atoms.

Why do atoms react?

Look at the three atoms in Fig 2.4.

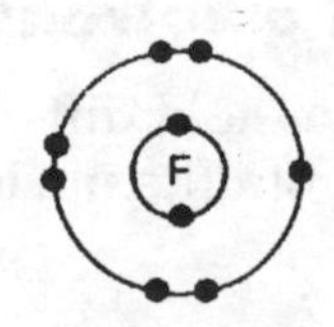

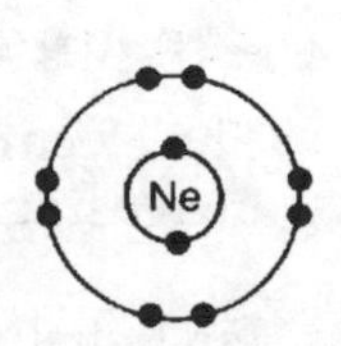

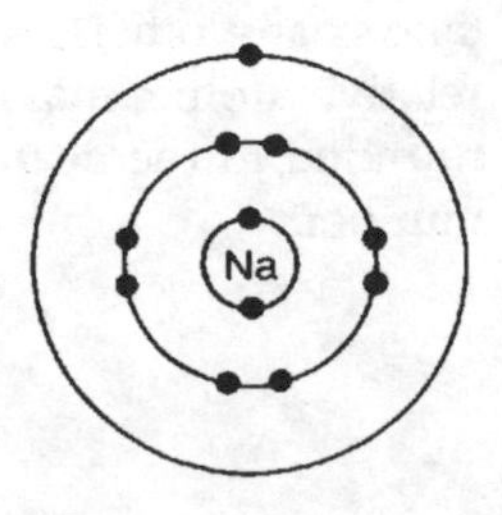

2.4 The electron arrangements of fluorine, F, neon, Ne and sodium, Na.

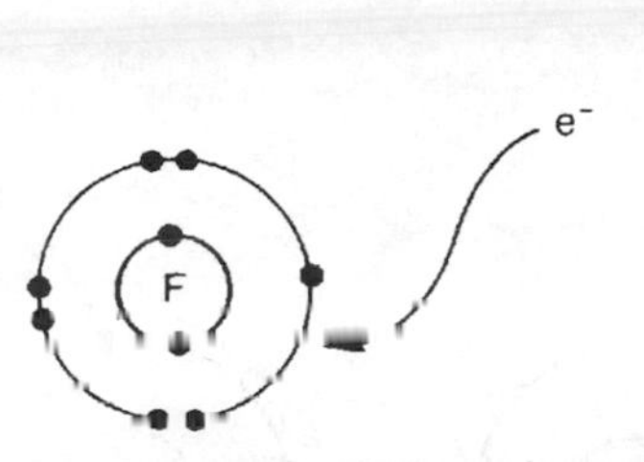

2.5 A fluorine atom gains an electron to get a full outer shell.

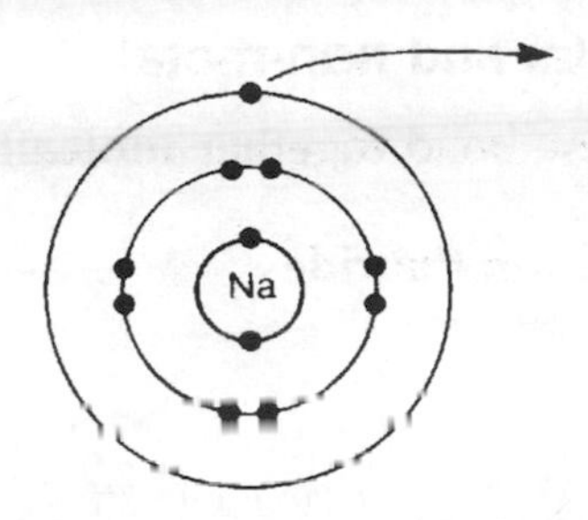

2.6 A sodium atom loses an electron (and its outer shell) to leave it with full electron shells.

The way in which they react is linked to their electron arrangements. Neon is unreactive. It has full shells of electrons.

Fluorine always reacts to get an extra electron. This gives it full electron shells (Fig 2.5).

Sodium always reacts to lose one electron. This gives it full electron shells (Fig 2.6).

When they react as shown, both fluorine and sodium become **ions**.

Ions

An ion is a charged atom (or group of atoms). Atoms that gain extra electrons become **negative ions**.

For example a fluorine atom (9 protons and 9 electrons) becomes a fluor*ide* ion, F^-, (9 protons and 10 electrons) and a sulphur atom (16 protons and 16 electrons) becomes a sulphide ion, S^{2-}, (16 protons and 18 electrons). **Note** negative ions change the ending of their names to -ide.

Atoms that lose electrons become **positive ions**. For example a sodium atom (11 protons and 11 electrons) becomes a sodium ion, Na^+, (11 protons and 10 electrons) and a magnesium atom (12 protons and 12 electrons) becomes a magnesium ion, Mg^{2+}, (12 protons and 10 electrons).

6 Neon atoms have 10 electrons (2,8) as do F^- ions and Na^+ ions. Why don't fluorine and sodium *become* Ne when they react?

7 Give the charge of the ions you would expect to be formed by a calcium, b oxygen, c aluminium, d nitrogen.

Electron arrangements of the first twenty elements

I	II	III	IV	V	VI	VII	0
H 1							**He** 2
Li 2, 1	**Be** 2, 2	**B** 2, 3	**C** 2, 4	**N** 2, 5	**O** 2, 6	**F** 2, 7	**Ne** 2, 8
Na 2, 8, 1	**Mg** 2, 8, 2	**Al** 2, 8, 3	**Si** 2, 8, 4	**P** 2, 8, 5	**S** 2, 8, 6	**Cl** 2, 8, 7	**Ar** 2, 8, 8
K 2, 8, 8, 1	**Ca** 2, 8, 8, 2						

Notice that the number of electrons in the outer shell is the same as the group number in the Periodic Table, except for Group 0 (which you could think of as having no electrons following a full outer shell).

Bonding

- All electrons are identical, but drawing some as dots and some as crosses helps us to see what is going on.

Where do all the electrons go to or come from? The answer is that atoms do not react singly, they need other atoms to react or bond with. We can keep track of what happens to electrons using **dot-cross diagrams**.

Metal and non-metal

These bond together **ionically**.

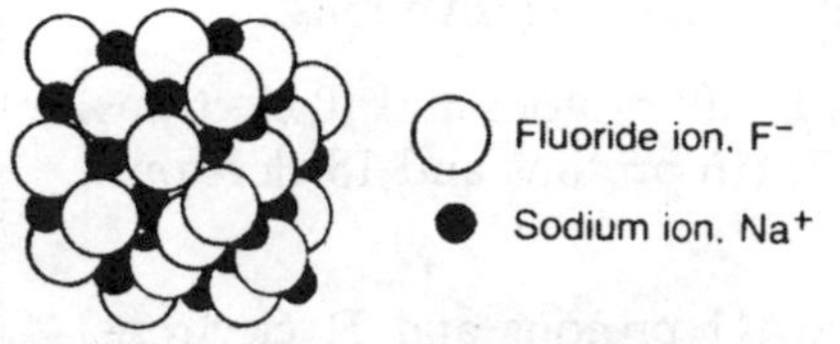

Sodium fluoride

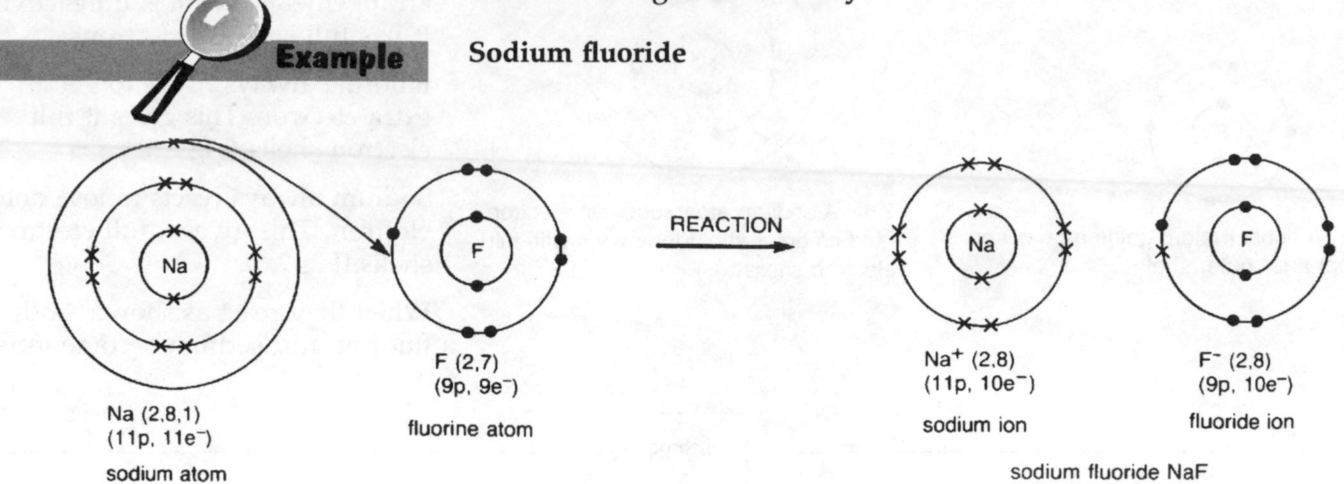

2.7 The formation of sodium fluoride.

2.8 The sodium and fluoride ions are regularly arranged in a giant structure.

Sodium needs to give away one electron to get a full outer shell and fluorine needs to gain one, so the formula is NaF (one to one).

Although the dot-cross diagram shows just one atom of each element, the ionic bonding extends throughout the structure. All the ions are held together in a **giant structure**. The bonding is strong and so the compound is hard to melt.

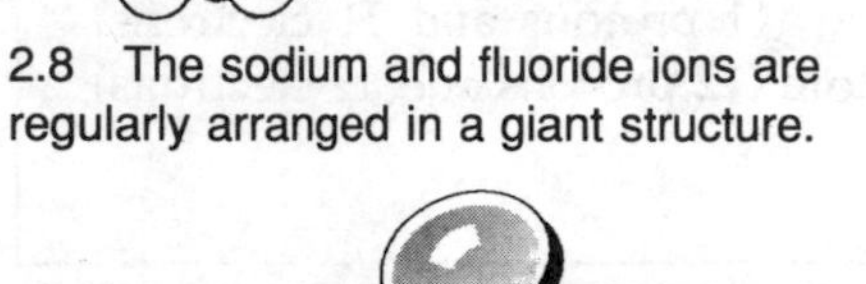

Magnesium fluoride

Magnesium needs to give away two electrons to get a full outer shell but fluorine needs to gain only one. So two fluorines are needed for each magnesium and the formula is MgF_2 (one to two). Magnesium fluoride also forms a giant structure.

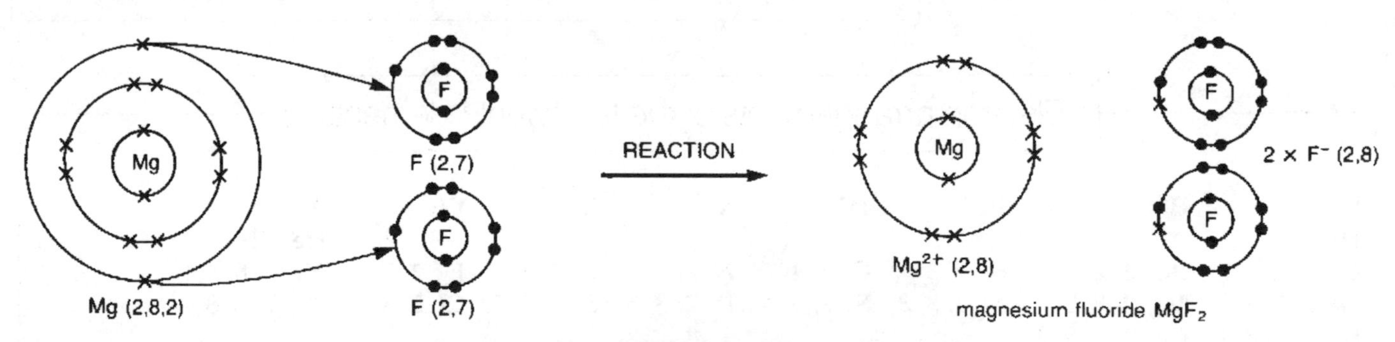

2.9 Dot-cross diagrams to show how magnesium fluoride is formed.

8 Draw similar diagrams to show the bonding between atoms of:

a sodium and chlorine
b sodium and oxygen
c calcium and fluorine

9 What is the formula and name of each of the products in question 8?

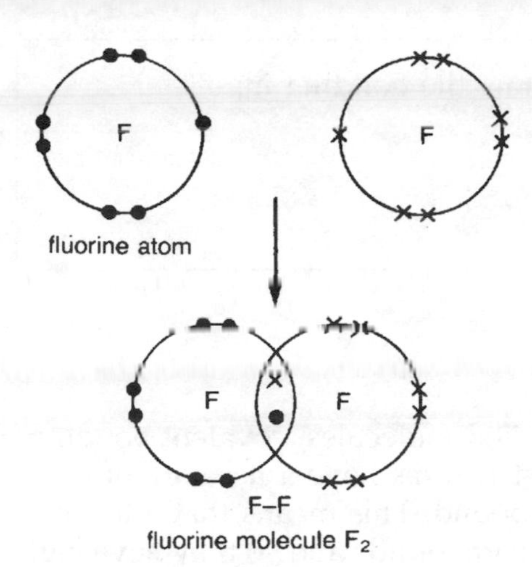

2.10 The bonding in fluorine gas F_2.

Properties of compounds with ionic bonding

- They are solids with high melting points.
- Ionic solids cannot conduct electricity as the ions are held fast.
- Melted ionic compounds can conduct electricity as the ions are now free to move.
- Many ionic compounds are soluble in water.
- If the compounds dissolve in water the solutions conduct electricity as the ions are now free to move.

Non-metal and non-metal

All non-metal atoms have (except for the noble gases) gaps in their outer shells, but in this case there are no metals to provide electrons to fill the gaps. Bonding between two non-metal atoms occurs by sharing of electrons. This is called **covalent** bonding. Small groups of atoms joined by covalent bonding are called **molecules.**

Two fluorine atoms

Each fluorine atom is bonded to just one other fluorine *within* the molecule. (See Fig 2.10.) In a sample of fluorine gas there would, of course, be vast numbers of other F_2 molecules, each one independent of the others. This molecular structure contrasts with the giant structures we have seen formed by ionic bonding.

Notice that:

It is usual to omit inner shells when dealing with covalent bonding.

- there is no transfer of electrons.
- there are no charged particles.
- the covalent bond is between only the atoms sharing the pair of electrons.
- there is no strong attraction *between* molecules.

Water (Fig 2.11) and oxygen (Fig 2.12)

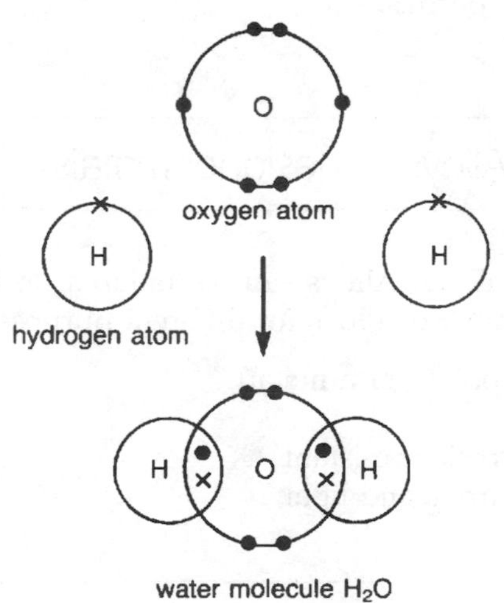

2.11 The bonding in water.

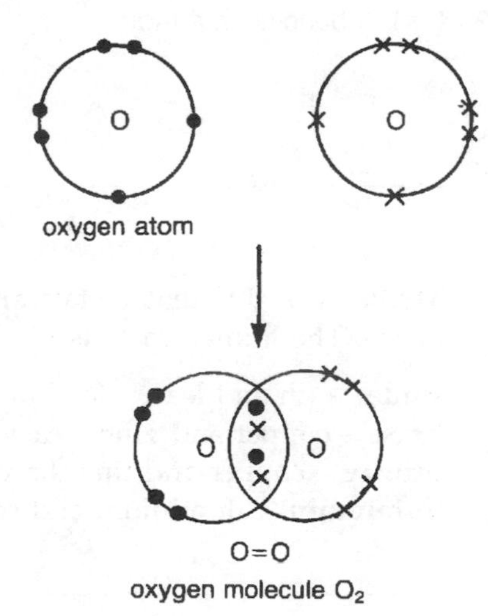

2.12 Oxygen molecules have a double bond.

Notice that the water molecule has two covalent bonds.

Notice that a total of four electrons is shared between the two oxygen atoms. This is called a double bond.

10 Draw dot-cross diagrams to show the bonding in:

a Chlorine, Cl_2
b Methane, CH_4
c Nitrogen, N_2

Giant covalent structures

As well as small groups of atoms called molecules, covalent bonding can produce giant structures. Such structures have a network of covalent bonds throughout the compound. This means that a lot of bonds have to be broken to melt the compounds and so they have high melting points. (See Fig 2.13.)

● silicon atom
○ oxygen atom
— represents a covalent bond, a pair of shared electrons

2.13 Silicon dioxide (the main component of sand) has a giant covalent structure.

Properties of covalently bonded compounds

Covalently bonded compounds, both molecular and giant:

- are poor conductors of electricity as there are no charged particles.
- often do not dissolve in water.
- if they do dissolve in water, their solutions are poor conductors of electricity.

Those with molecular structures are gases, liquids or low melting point solids. Those with giant structures are high melting point solids.

Metals

All metal atoms have one, two or three electrons in their outer shells. They can bond to other metal atoms by contributing these electrons to a general pool of electrons that are free to move throughout the resulting giant structure. See Fig 2.14. This is called metallic bonding

It is the mobile electrons that give metals and alloys their unique properties.

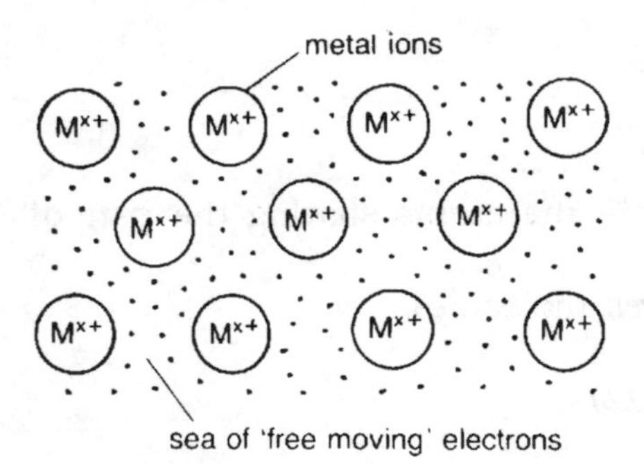

2.14 The bonding in a metal.

Alloys – 'designer metals'

Mixtures of different metals are called alloys. Alloys can be 'tailored' to match their properties to particular needs. The human race uses a large number of alloys for different purposes. Some examples include

solder – tin and lead – for low melting point (for a metal)
brass – copper and zinc – easy to work
bronze – copper and tin – hard and corrosion resistant
duralumin – aluminium and copper – strong and light

Properties of substances with metallic bonding

- They have giant structures with high melting points.
- They conduct electricity as solids or liquids.
- They are shiny, malleable (can be shaped by beating) and ductile (can be pulled into wires).

11 Which is the only metal element that does not have a giant structure at room temperature?

Structure and bonding – Summary

BONDING describes what holds atoms together. There are three types which occur in different situations and result in different properties.

Ionic bonding occurs by electron transfer from a metal to a non-metal. The resulting compounds conduct electricity when liquid or in solution but not when solid.

Covalent bonding occurs by electron-sharing between non-metal atoms. The resulting compounds do not conduct electricity in any physical state.

Metallic bonding occurs by pooling of electrons between metal atoms. Metals conduct both as liquids and solids.

STRUCTURE describes the arrangement of atoms. There are two main types – giant and molecular.

Giant structures have a regular 3-D arrangement of atoms which extends throughout the compound. The structure may be held together by any of the three types of bonding. Giant structures have high melting and boiling points.

Molecular structures consist of many small units (molecules) which each consist of a few atoms held together by covalent bonds. The molecules are not strongly bonded to one another although the bonding *within* them is strong. Molecular structures can be recognised by their low melting and boiling points.

12 Table 2.2 below gives some information about four substances. State which one is ionic, which metallic, which has a molecular structure and which has a covalent giant structure.

Table 2.2

Substance	Melting point/ °C	Boiling point/ °C	Electrical conduction	
			As solid	As liquid
A	1083	2567	good	good
B	−182	−164	poor	poor
C	1723	2230	poor	good
D	993	1695	poor	poor

Some glimpses ahead

X-ray diffraction

The method of X-ray diffraction provides direct evidence for some of our theories of bonding. This technique allows us to see where the electrons are in a molecule or giant structure. The electrons show up rather like contour lines on a map.

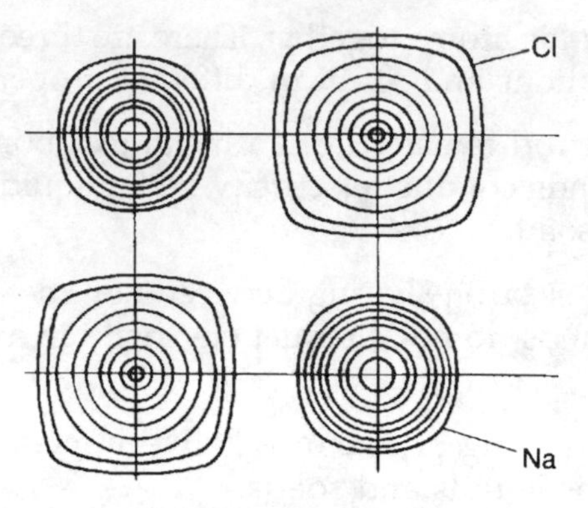

2.15 An electron density map of sodium chloride, NaCl, an ionic compound.

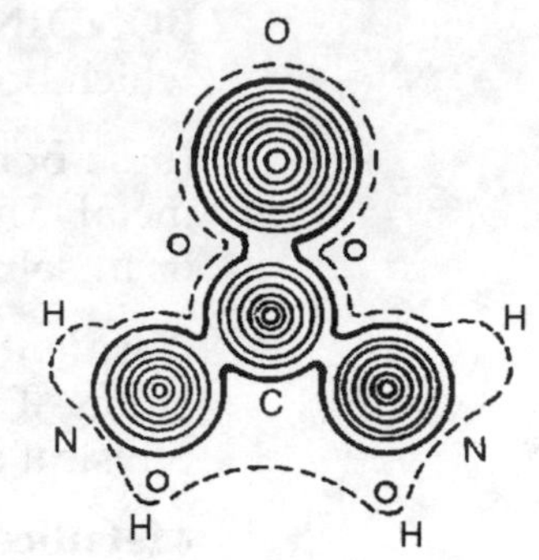

2.16 An electron density map for urea, $CO(NH_2)_2$, a covalent compound.

Fig 2.15 represents an ionic compound as the separate ions are obvious while Fig 2.16 is a covalent compound as the sharing of electrons can be seen.

The bonding of HCl

The dot-cross diagrams in Fig 2.17 show two ways in which an atom of hydrogen could bond to one of chlorine – one ionic and one covalent. Which is correct? The answer seems to be both.

Pure hydrogen chloride is a gas which does not conduct electricity – just what we would expect for a covalently bonded molecular compound. However, when we dissolve it in water, hydrogen chloride does conduct electricity – as we would expect for an ionic compound. The truth seems to be that the bonding in this substance (and many others) is somewhere between the two extremes.

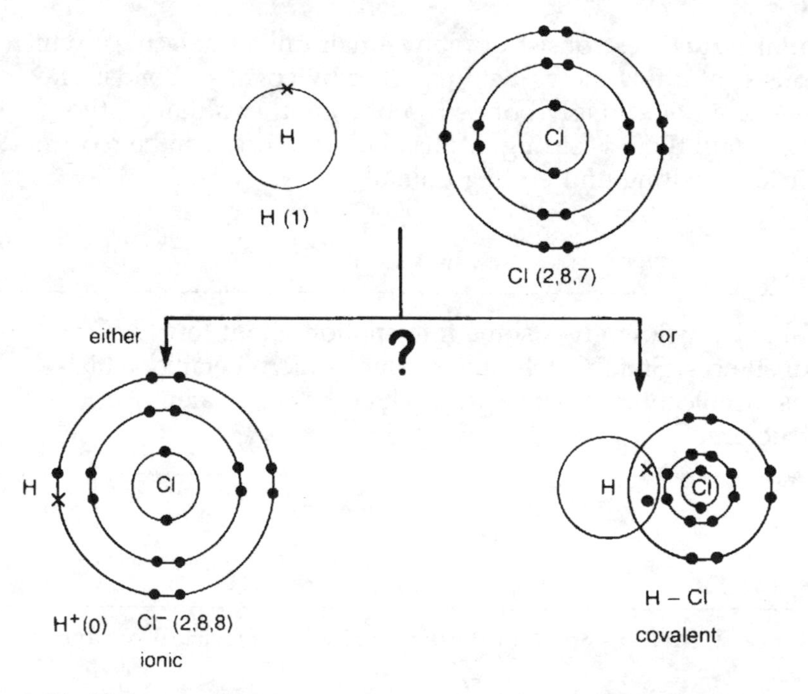

2.17 Hydrogen and chlorine could be bonded covalently or ionically.

Chapter 3 Formulae

STARTING POINTS

- Do you understand the following terms: **atom, element, compound, molecule, ion, giant structure, molecular structure, covalent bonding?**

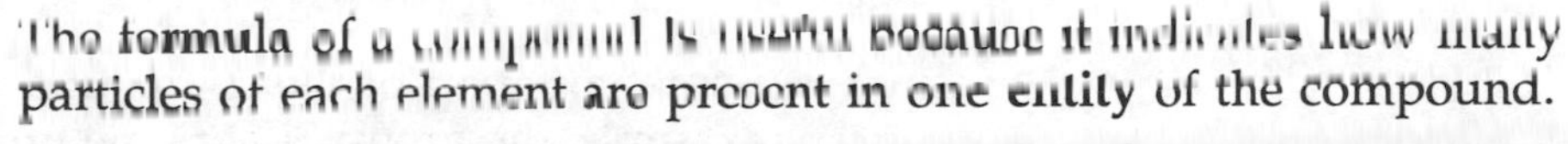

The formula of a compound is useful because it indicates how many particles of each element are present in one entity of the compound.

The word entity is used to describe the simplest formula unit of a compound. It can be used to refer to compounds with molecular structures such as water, H_2O, or ammonia, NH_3 or giant ionic such as sodium chloride, NaCl, or calcium chloride, $CaCl_2$.

CaS	means one atom of calcium to one atom of sulphur.
$CaCl_2$	means one atom of calcium to two atoms of chlorine.
H_2SO_4	means two atoms of hydrogen to one atom of sulphur to four atoms of oxygen.

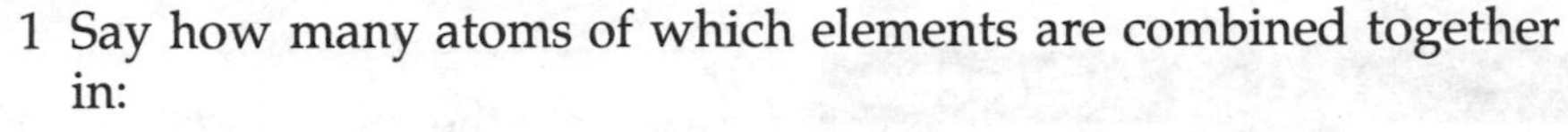

1 Say how many atoms of which elements are combined together in:

a KF b NaCl c C_2H_6 d $C_6H_5NO_2$

Note Chemical symbols normally have one or two letters. In two-letter symbols, the first letter is always a capital and the second letter small. So Co is the symbol for cobalt, while CO stands for one atom of carbon and one of oxygen. You will need to use the Periodic Table (inside back cover) to look up the element that each symbol represents.

Brackets

Sometimes formulae are written using brackets. A small number after a bracket multiplies all the elements in the bracket by that number.

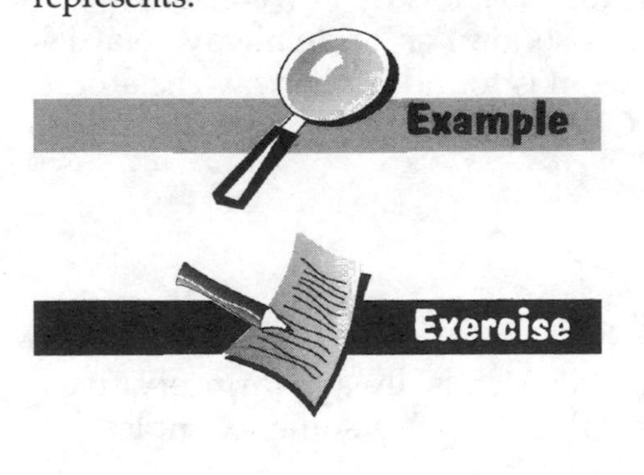

$Ca(OH)_2$: each entity contains 1 calcium, 2 oxygen and 2 hydrogen atoms
$Ca(NO_3)_2$: each entity contains 1 calcium, 2 nitrogen and 6 oxygen atoms

2 Say how many atoms of each element are present in:

a $(NH_4)_2SO_4$ b $Mg(NO_3)_2$ c $Ca_3(PO_4)_2$

Naming compounds

When the compound consists of a metal and a non-metal the naming is straightforward, we put the names of the two elements together, metal first, and change the ending of the non-metal to **-ide**. This ending signifies that there are only two elements in the compound.

	(METAL NAME)	(NON-METAL NAME) ('IDE')
NaCl	Sodium	chloride
KI	Potassium	iodide
$CaCl_2$	Calcium	chloride

3 Name the following:

a $LiBr$ b Na_2O c MgO
d MgI_2 e FeS f Al_2O_3

Sometimes part of the compound consists of a group of particles. For example, the group SO_4. Some groups of atoms are given special names. Ones you will often meet are

SO_4	sulphate	These behave as non-metals and form the second part of the name.
NO_3	nitrate	
CO_3	carbonate	
PO_4	phosphate	
OH	hydroxide	
NH_4	ammonium	This behaves as a metal and forms the first part of the name.

From these examples you may be able to see that the ending **-ate** is often used to mean 'and oxygen as well' although hydroxide is an exception.

4 Name the following:

a Li_2CO_3 b $Ca(OH)_2$ c $AlPO_4$
d $MgSO_4$ e $Cu(NO_3)_2$ f NH_4Cl

3.1 Which compound named in question 4 might be relevant to this policeman?

Formulae and bonding

Why do compounds have constant formulae? Why is water always H_2O rather than HO or HO_2? Why is sodium chloride always $NaCl$ rather than Na_2Cl or $NaCl_2$? The reason is found in the way the atoms in the compound are bonded (see Chapter 2).

Ionic compounds

These contain a metal and a non-metal. Metals always form positive ions and non-metals negative ions. Table 3.1 gives some examples.

Chapter 2 explains how we can predict the charges on some of these ions.

Since all compounds are neutral, we can predict the formula of an ionic compound. The total positive charge must be the same as the total negative charge.

Copper sulphate consists of two types of ion, copper Cu^{2+} and sulphate SO_4^{2-}. The compound resulting from these particles must be neutral overall and each Cu^{2+} is balanced by one SO_4^{2-}. So the formula is $CuSO_4$. In aluminium oxide, aluminium forms Al^{3+} and oxygen O^{2-}. The simplest way to get a neutral compound is if two aluminium ions (total charge 6+) combine with three oxide ions (total charge 6–), so the formula is Al_2O_3.

Table 3.1 The charges on some ions

Positive ions			Negative ions		
1+	2+	3+	1−	2−	3−
Li^+ lithium	Mg^{2+} magnesium	Al^{3+} aluminium	Cl^- chloride	O^{2-} oxide	PO_4^{3-} phosphate
Na^+ sodium	Ca^{2+} calcium		Br^- bromide	S^{2-} sulphide	
K^+ potassium	Cu^{2+} copper		I^- iodide	CO_3^{2-} carbonate	
NH_4^+ ammonium	Zn^{2+} zinc		NO_3^- nitrate	SO_4^{2-} sulphate	
			OH^- hydroxide		

5 Use Table 3.1 above to predict the formulae of the following compounds:

a lithium carbonate
b aluminium iodide
c sodium nitrate
d zinc sulphate
e ammonium phosphate
f calcium oxide
g copper chloride
h aluminium phosphate

Metal ions

- Metal ions are always positive.
- Metals in groups I, II and III have the same charge as their group number when they form ions.
- Transition metal atoms can often form two or more ions with different charges. Some common transition metal ions are given in Table 3.2.

Table 3.2 Some transition metal ions

1+	2+	3+
Ag^+	Cu^{2+}	Cr^{3+}
Au^+	Mn^{2+}	Fe^{3+}
Cu^+	Fe^{2+}	

Non-metal ions

- Non-metal elements in group VII always have a charge of 1− when they bond ionically.
- Non-metal elements in group VI always have a charge of 2− when they bond ionically.
- Some non-metal ions consist of more than one atom. The charges of some of these complex non-metal ions are given in Table 3.1.

Covalently bonded compounds

These are bonded together by electron-sharing. How many other atoms a particular atom can bond with depends on how many electrons it needs to gain a full outer shell, see Chapter 2.

Oxygen with six electrons in its outer shell needs two more, so it could bond with two hydrogen atoms which each need one extra electron to form water, H_2O. Each oxygen forms two **single bonds** (two electrons shared). See Fig 3.2.

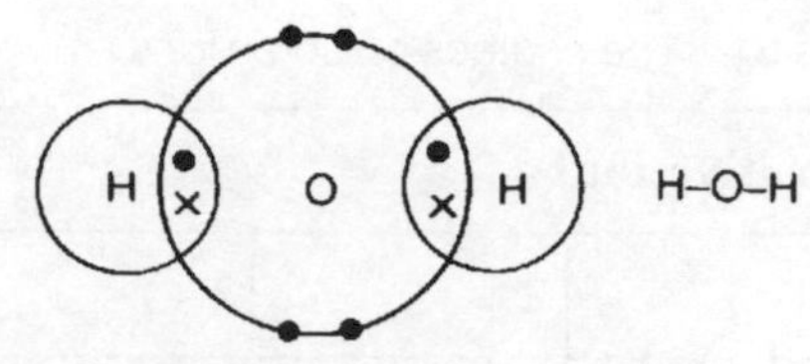

3.2 The bonding in water, H_2O. Oxygen forms two single covalent bonds.

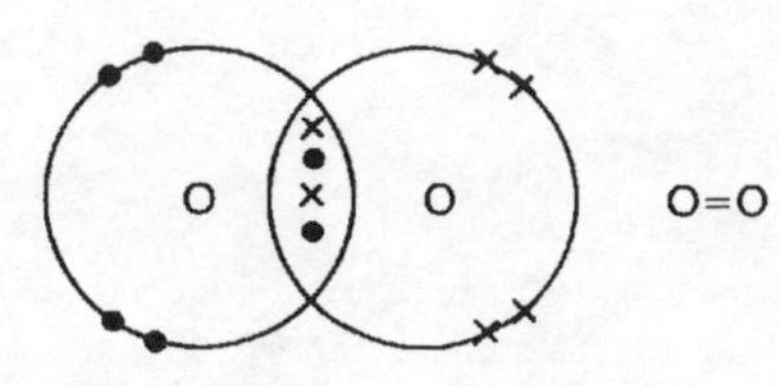

3.3 Oxygen can form two covalent bonds to make a double bond.

Alternatively it could bond with another oxygen so that both atoms share a total of four electrons. This is a **double bond**. See Fig 3.3.

Combining power (valency)

Because it can form only one single bond, hydrogen is said to have a combining power or **valency** of one. Oxygen, which can form either two single bonds or one double bond, has a combining power of two.

6 Use the Periodic Table to help you work out the electron arrangement of the following atoms and use it to predict the combining power of each.

a Nitrogen, N
b Carbon, C
c Chlorine, Cl
d Sulphur, S
e Argon, Ar
f Silicon, Si

What do you notice about the combining power of carbon and silicon which are in the same group?

Some formulae

Here are the formulae of some interesting compounds.

aspirin $C_9H_8O_4$

TNT $C_7H_5N_3O_6$

insulin (the hormone injected by diabetics to control their blood sugar levels) $C_{254}H_{377}N_{65}O_{75}S_6$

ethyl mercaptan (according to the Guinness Book of Records, the most evil-smelling compound known – its smell is a cross between rotting cabbage, garlic, onions and sewer gas) C_2H_5SH

ethanol (the alcohol in alcoholic drinks) C_2H_5OH

Notice how similar the formulae of the last two are – what a difference an atom makes!

3.4 'My mouth tastes like C_2H_5SH, I've had too much C_2H_5OH – I need some $C_9H_8O_4$.'

Shapes of molecules

Formulae can tell us much more than just the numbers of different types of atoms in a molecule. They can tell us which atoms are bonded to which and even the shape of the molecule. Shape has a great deal to do with how molecules like drugs, for example, work. Chemists can now tailor compounds to the shape they want – designer molecules! They need to use computer graphics like the ones below to record the formulae.

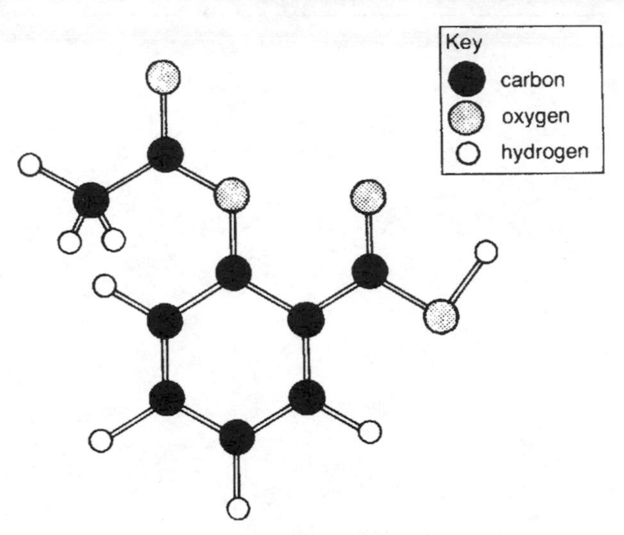

3.5 Computer graphic of aspirin, one of the oldest drugs known, but it is still finding new uses.

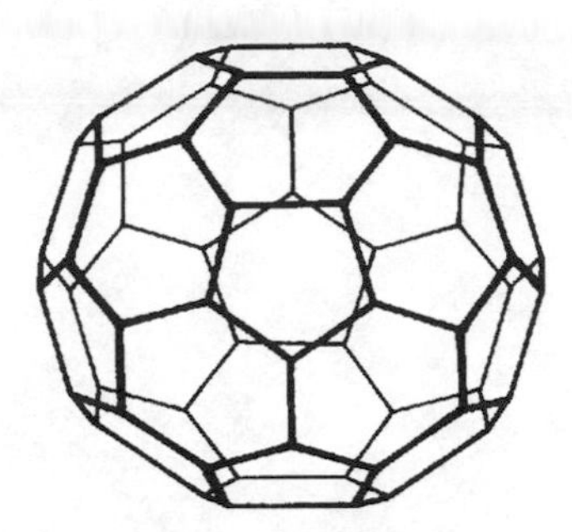

3.6 Buckminsterfullerene – a molecular football, C_{60}, which consists entirely of carbon atoms. Drawing it in this way is called skeletal notation and one carbon atom is implied at every junction.

Bonding within the nitrate ion

The nitrate ion, NO_3^-, is an ion but it consists of four atoms covalently bonded together. Here are some facts about it:

- it is composed of one nitrogen and three oxygens
- nitrogen and oxygen are both non-metals and therefore the bonding within the ion is covalent
- the arrangement of the particles is such that nitrogen is surrounded by the three oxygens
- nitrogen has five electrons in its outer shell and oxygen has six.

Fig 3.7 shows a dot-cross diagram (see Chapter 2) for the covalent bonding within the nitrate ion.

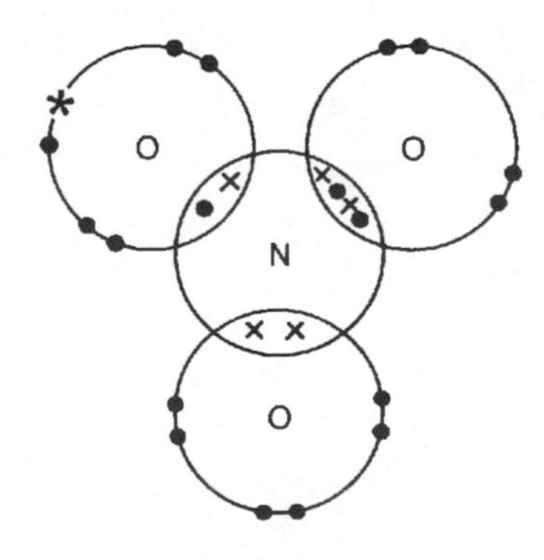

3.7 The nitrate ion, NO_3^-.

Notice
- All the atoms now have full outer electron shells.
- One oxygen has formed a double bond with the nitrogen (symbol =).
- One oxygen has formed a covalent bond with the nitrogen in which both the shared electrons came originally from the nitrogen atom. This is called a dative bond (symbol →, showing which atom has donated the pair of electrons to the bond).
- If you count the electrons, you will find that there is one more than we started with. This has come from the metal atom which goes with the nitrate ion. This electron is marked * in the diagram. One extra electron is needed to allow each atom to have a full outer shell of electrons.

In fact this is not the whole story about bonding in NO_3^- as you will find in your advanced course.

7 Draw dot-cross diagrams like the one in Fig 3.7 for the nitrate ion for the following ions:

a the hydroxide ion OH^-.

b the carbonate ion CO_3^{2-}.

This is similar in structure to the nitrate ion but carbon has only four outer shell electrons.

Chapter 4 Equations

Word and symbol equations

Word equations tell us which chemicals will react together and what products they form. Balanced symbol equations have a different function. They tell us in what proportions substances react together.

STARTING POINTS

- You should know that equations represent the reactions that happen in real life. They are found by experiment. This means that just because you can write an equation, it doesn't follow that a reaction will actually happen.
- You should recognise the chemical symbols for the most commonly used elements and know how to look up others in the Periodic Table.

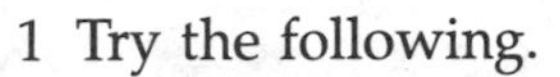

1 Try the following.

a Write a word equation for what happens when magnesium is burned in oxygen.

b Say what the following state symbols mean in an equation – (s), (l), (g), (aq)

4.1 Magnesium burning in air. A little of the magnesium will also combine with the nitrogen in air.

2 Can you recognise a **balanced equation**? Which of the following equations is/are balanced?

a $2H_2 + O_2 \rightarrow 2H_2O$

b $CH_4 + O_2 \rightarrow CO_2 + H_2O$

c $Na + H_2O \rightarrow 2NaOH + H_2$

d $H_2 + Cl_2 \rightarrow 2HCl$

A balanced equation has the same number of each type of atom on each side of the arrow, because chemical reactions simply rearrange atoms, they do not create or destroy them.

Although equations can only be found for certain by experiment, we can often work out what the balanced equation for a given reaction must be if we know the reactants (starting materials), the products and their formulae.

3 Balance the two equations below. (Don't change the formulae!)

a $Li + F_2 \rightarrow LiF$

b $Fe + O_2 \rightarrow Fe_2O_3$

Check your answers to question 3 in the answer section. If you were right, and feel confident about balancing equations, move on to the **Types of reactions** section. Otherwise read the section that follows on balancing equations.

Balancing equations – introduction

Suppose we want to balance the equation

$$Na + Cl_2 \rightarrow NaCl$$

The problem with the equation above is that there are two chlorine atoms on the left-hand side and only one on the right-hand side.

Step 1 To get two chlorines on the right-hand side put a 2 in front of NaCl

$$Na + \overset{\checkmark}{Cl_2} \rightarrow \overset{\checkmark}{2\,NaCl}$$

But now the chlorine is correct and not the sodium.

Step 2. Go back to the left-hand side and put a 2 in front of the Na.

$$\overset{\checkmark}{2\,Na} + \overset{\checkmark}{Cl_2} \rightarrow \overset{\checkmark\checkmark}{2\,NaCl} \quad - \text{ balanced}$$

Balancing equations

Balancing an equation is a bit like the problem of buying batteries for a torch. If the batteries come in packs of two and the torches take three batteries each, what is the smallest number of items to buy to get a set of working torches with nothing left over? A little thought should convince you that if you buy three packs of batteries and two torches you can assemble two working torches.

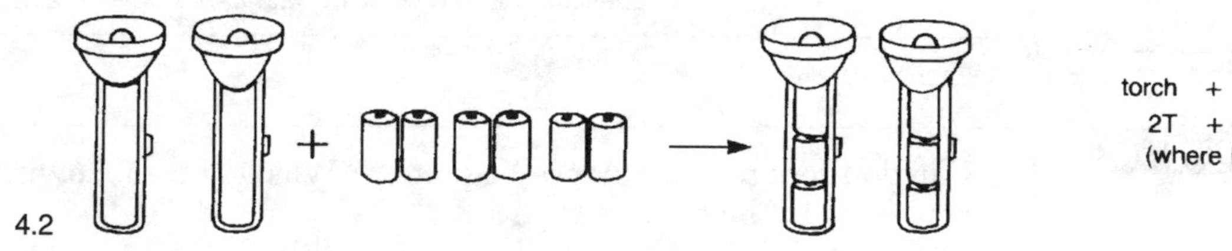

4.2

torch + batteries → working torch

$2T + 3B_2 \rightarrow 2TB_3$

(where B = 1 battery, T = 1 torch)

This is just like balancing the equation for the reaction of aluminium (Al) with chlorine (Cl) to give aluminium chloride ($AlCl_3$).

$$2\,Al + 3\,Cl_2 \rightarrow 2\,AlCl_3$$

Hints for balancing equations

Remember that a balanced equation must have the same number of atoms of each element on each side of the equation. Here are a few hints.

- You must not change the formulae – these have been found by experiment and cannot be changed at your convenience.
- You can only change the number of atoms by putting numbers *in front of* the formulae.
- You often need more than one step to get it right.
- If you can't balance the equation in 3–4 steps, you probably need to check that you have all the formulae correct.

4 Balance the following equations

a $Mg + HCl \rightarrow MgCl_2 + H_2$
b $Na + O_2 \rightarrow Na_2O$
c $Ca(OH)_2 + HNO_3 \rightarrow Ca(NO_3)_2 + H_2O$
d $Ca + H_2O \rightarrow Ca(OH)_2 + H_2$

Types of reactions

Atoms in chemical substances can rearrange in all sorts of ways when they react but there are groups of reactions which follow the same patterns and which have their own names.

Direct combination

Did you know that popping hydrogen is a direct combination reaction?

Writing the equation for two elements reacting together is probably the easiest one to do. As you have seen in the chapter on formulae, the name of a compound of two elements always ends in -ide, so the word equation is straightforward. The difficult part is getting the formulae of the resulting compounds correct. You may be able to look them up or work them out – see Chapter 3.

5 Write the word equation and the the balanced symbol equations for the following:

a iron reacting with sulphur
b magnesium reacting with nitrogen
c sodium reacting with bromine

(The formulae you need to use are NaBr, Mg_3N_2, FeS.)

Oxidation and reduction

Oxidation

This, at its simplest, is what happens when we burn substances in air thus adding oxygen to them.

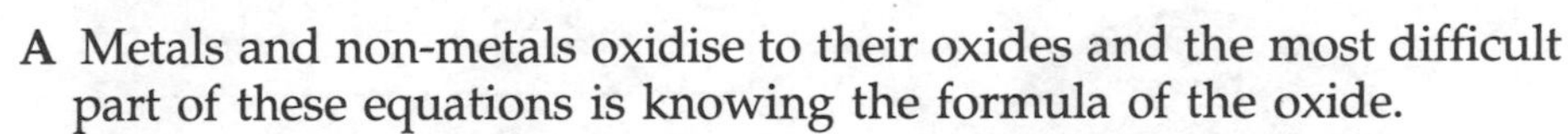

A Metals and non-metals oxidise to their oxides and the most difficult part of these equations is knowing the formula of the oxide.

copper + oxygen → copper oxide
$2Cu + O_2 \rightarrow 2CuO$

B Or we can have compounds which already contain oxygen reacting to get more.

$2SO_2 + O_2 \rightarrow 2SO_3$

Sulphur dioxide has been oxidised to sulphur trioxide.

6 Write the equations for the oxidation to their oxides of

a magnesium, b lithium, c aluminium,
d carbon, e hydrogen.

See if you can work out the formulae of the oxides without looking at the line overleaf. See Chapter 3.

Formulae [Li_2O, CO_2, Al_2O_3, MgO, H_2O]

Reduction

This, at its simplest, means removal of oxygen.

Hot copper oxide can be reduced to copper using hydrogen gas.

$$CuO + H_2 \rightarrow Cu + H_2O$$

Notice that the hydrogen has been oxidised in this reaction.

Redox reactions

The definition of oxidation and reduction has been extended to include other reactions which may not include oxygen at all!

Look at the reaction below.

$$2Cu + O_2 \rightarrow 2CuO$$

The copper is neutral at the beginning of the reaction, but has formed a Cu^{2+} ion when it reacts with oxygen. **It has lost electrons.**

Oxygen is neutral at the start of the reaction and it forms an oxide ion O^{2-} when it reacts with copper. **It has gained electrons.**

The extended definition of oxidation reduction reaction is:

Oxidation is loss of electrons
Reduction is gain of electrons (Use the phrase OIL RIG to recall this.)

and the general term is **redox reactions.**

Copper has been oxidised and oxygen reduced in the above example.

It is always the case that whenever an element is oxidised another is reduced in the same reaction.

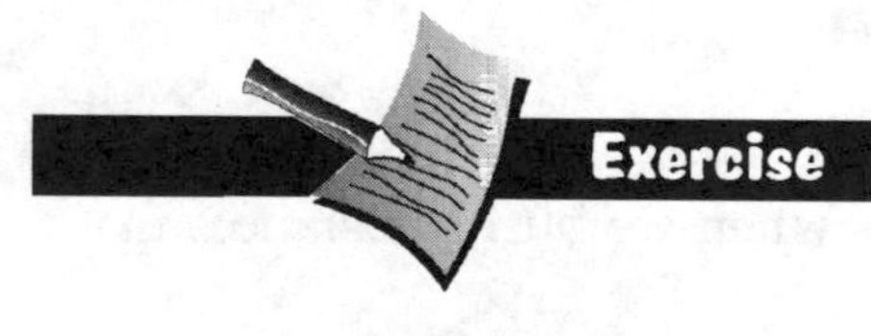

7 What is happening to sodium and chlorine in this reaction and why is it a redox reaction?

$$2Na + Cl_2 \rightarrow 2NaCl$$

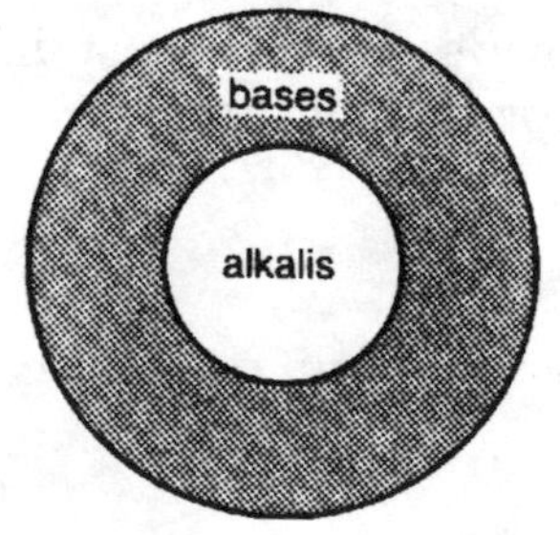

4.3 An alkali is a soluble base.

A base is a metal oxide, such as copper oxide, or hydroxide, such as sodium hydroxide, (although ammonium hydroxide, NH_4OH is included because the ammonium ion NH_4^+ behaves like a metal ion). Bases will neutralise acids. A base that dissolves in water, such as potassium hydroxide, is an alkali.

Neutralisation

Another group of reactions is that of the **acids**. Acids form **salts** when they react, so you need to know the names of the salts that are produced by each acid. Here are some common ones.

Acid	Formula	Type of salt	Example of salt
sulphuric	H_2SO_4	sulphate	Na_2SO_4
hydrochloric	HCl	chloride	KCl
nitric	HNO_3	nitrate	$Pb(NO_3)_2$
ethanoic	CH_3COOH	ethanoate	$LiCH_3COO$

The main reactions are:

acid + reactive metal → salt + hydrogen
acid + base → salt + water
acid + metal carbonate → salt + water + carbon dioxide

Ionic equations

When an acid reacts with an alkali in solution, we can write the equation in terms of ions. For example, hydrochloric acid reacts with sodium hydroxide to give sodium chloride and water:

$$HCl(aq) \quad + \quad NaOH(aq) \quad \rightarrow \quad NaCl(aq) \quad + \quad H_2O(l)$$

When you do this experiment, you start with two colourless solutions neither of which is neutral, and end up with a neutral colourless solution.

We can write the equation in terms of the ions in the solutions and see what is happening.

$HCl(aq)$ exists as ions $H^+(aq)$ and $Cl^-(aq)$

$NaOH(aq)$ exists as ions $Na^+(aq)$ and $OH^-(aq)$

One product of the reaction, sodium chloride, exists as ions $Na^+(aq)$ and $Cl^-(aq)$.

So we can write the reaction:

$$H^+(aq) + Cl^-(aq) + Na^+(aq) + OH^-(aq) \rightarrow Cl^-(aq) + Na^+(aq) + H_2O(l)$$

Overall what has happened is:

$$\mathbf{H^+(aq) \quad + \quad OH^-(aq) \quad \rightarrow \quad H_2O(l)}$$

The other ions have not changed. They are called spectator ions.

Whenever an acid reacts with an alkali, the overall reaction will be the same as the one above.

4.4 'Aren't you glad to be a spectator?'

8 a Magnesium and sulphuric acid react to give magnesium sulphate and hydrogen. Write (or say to yourself) the word equations for the reaction between each of the acids nitric, hydrochloric and ethanoic with magnesium. Then write balanced equations for the reactions. Pick out the formulae you will need: $Mg(NO_3)_2$, $MgSO_4$, $Mg(CH_3COO)_2$, $MgCl_2$.

b Do the same thing as above for the reaction between sodium hydroxide, $NaOH$, and the four acids. Pick out the formulae you will need: Na_2SO_4, $NaCH_3COO$, $NaNO_3$, $NaCl$.

c Write balanced symbol equations for the reactions between sodium carbonate and the four acids given in (b) and then magnesium carbonate and the four acids. The only new information you need is the formulae of sodium carbonate Na_2CO_3 and magnesium carbonate $MgCO_3$.

If you can confidently write and balance the equations in question 8, then you can extend this to any Group I or Group II metal in the reacting substance.

Double decomposition

The solutions of two different salts can react together and form an insoluble salt. If you were mixing the solutions together in a test tube you would see a solid appearing in the test tube. This is called a **precipitate**.

In the symbol equation, (aq) means that the salt is dissolved in water, and (s) would be the precipitate.

$$CuSO_4(aq) + Na_2CO_3(aq) \rightarrow CuCO_3(s) + Na_2SO_4(aq)$$

The ions have 'swapped partners'. This is called **double decomposition**.

9 Describe what you would see happening in the above reaction.

10 Sodium and potassium salts and all nitrates are always soluble in water. Use this information to write word and symbol equations for the following double decomposition reactions which take place in solution. Include state symbols in your answers.

a lead nitrate and potassium iodide
b potassium carbonate and barium chloride
c sodium chloride and silver nitrate

Pick out the formulae you will need: $BaCl_2$, KI, $Pb(NO_3)_2$, K_2CO_3, NaCl, $AgNO_3$, $NaNO_3$, AgCl, PbI_2, KCl, $BaCO_3$, KNO_3

Reversible reactions

Most of the reactions that we are familiar with seem to start with the reactants and end up only with the products. We could say that everything about the reaction drives it from left to right.

Reactants $\rightarrow$ Products

For example

magnesium + copper sulphate $\rightarrow$ magnesium sulphate + copper

$$Mg(s) + CuSO_4(aq) \rightarrow MgSO_4(aq) + Cu(s)$$

If we try reacting copper with magnesium sulphate nothing happens.

11 Explain why copper and magnesium sulphate do not react.

However, with some reactions the reverse reaction *does* take place and as soon as the products form they react to produce the reactants again.

The end result is a mixture of reactants and products.

We call these reactions reversible, and we say that the final mixture has reached equilibrium (a balance point). The symbol $\rightleftharpoons$ is used to show a reversible reaction. Both the forward and back reactions are continually taking place.

Reactants $\rightleftharpoons$ Products

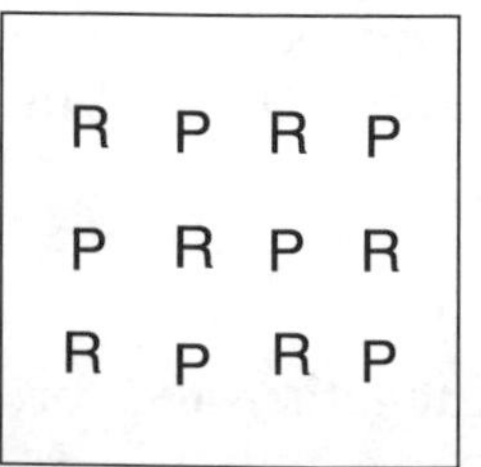

4.5 An equilibrium mixture of reactants (R) and products (P).

This reversibility is a nuisance in industrial processes because only the product is wanted. However, the proportion of reactants to products in the final equilibrium mixture can be changed by changing the conditions under which the reaction takes place. It can be affected by pressure, temperature and concentration.

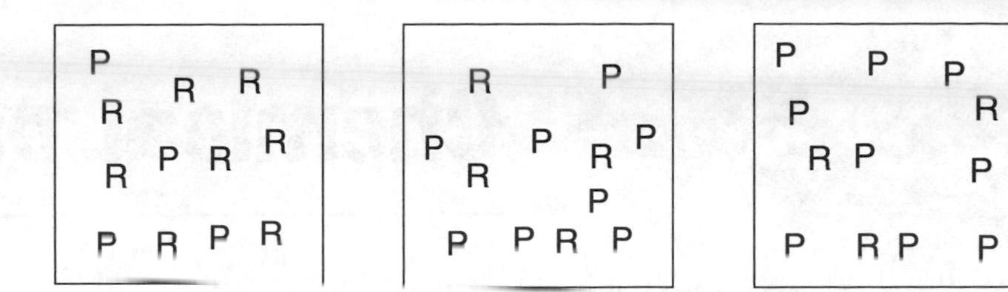

4.6 The proportions of reactants to products can vary in an equilibrium mixture.

The industrial production of ammonia

Ammonia is a vital industrial chemical which is needed in the manufacture of fertilisers, dyes, drugs and explosives. Its formation from nitrogen and hydrogen is a reversible reaction:

nitrogen + hydrogen ⇌ ammonia

$$N_2(g) + 3\,H_2(g) \rightleftharpoons 2\,NH_3(g)$$

The graphs show the percentage of ammonia in the equilibrium mixture under various conditions of pressure and temperature.

Compromise conditions are used in the industrial process because a low temperature slows down the reaction, and the higher the pressure the more expensive the plant is to build.

An iron-based catalyst speeds up the reaction but does not change the final proportion of reactants to products as catalysts speed up the forward and back reactions equally.

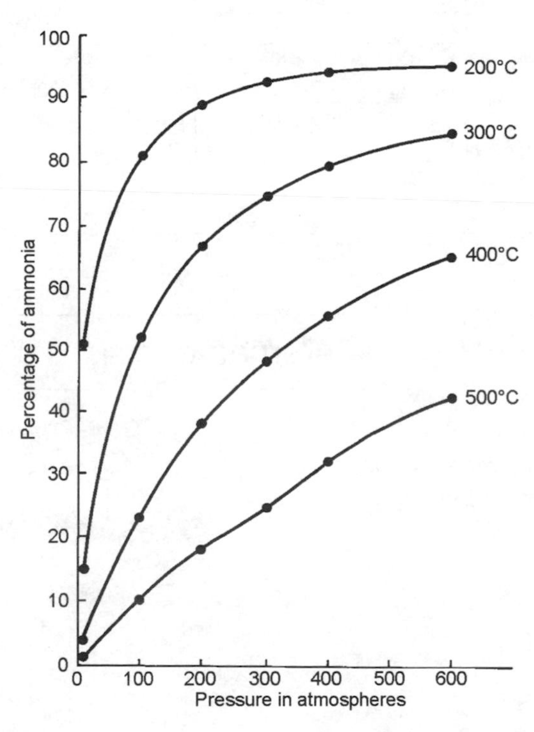

4.7 The effect of pressure and temperature on the percentage conversion of hydrogen and nitrogen to ammonia.

12 Look at the graph of ammonia production (Fig 4.7).

a What is the effect of increasing the temperature on the percentage of ammonia in the equilibrium mixture?

b What is the effect of increasing the pressure on the percentage of ammonia in the equilibrium mixture?

Chapter 5 Chemical patterns

STARTING POINTS

- Do you understand the following terms: **element, compound, relative atomic mass?**

The Periodic Table

This is a chemical map showing all the elements in order of their **atomic number,** Z. It was first drawn up by the Russian **Dimitri Mendeleev** in 1869.

The work of Mendeleev

Mendeleev thought that if his table of the elements was to be really useful, it should be able to *predict* new information, not just explain what was already known. His original arrangement is shown below.

Mendeleev arranged all 63 elements then known in rows and columns in order of their relative atomic masses. However he realised that there might well be undiscovered elements and left gaps for them in sensible places. He predicted the properties of these unknown elements by averaging the properties of the elements on either side of the gap. The elements from the gaps have now all been discovered and fit Mendeleev's predictions extremely closely.

H							
Li	Be	B	C	N	O		
Na	Mg	Al	Si	P	S	Cl	
K Cu	Ca Zn		Ti	V As	Cr Se	Mn Br	Fe Co Ni
Rb Ag	Sr Cd	Y In	Zr Sn	Nb Sb	Mo Te	I	Ru Rh Pd
Cs Au	Ba Hg	La Tl	Pb	Ta Bi	W		Os Ir Pt
I	II	III	IV	V	VI	VII	VIII

5.1 Mendeleev's Periodic Table.

Alternative versions of the Periodic Table

There are various other possible arrangements as well as the familiar modern one shown on page 96.

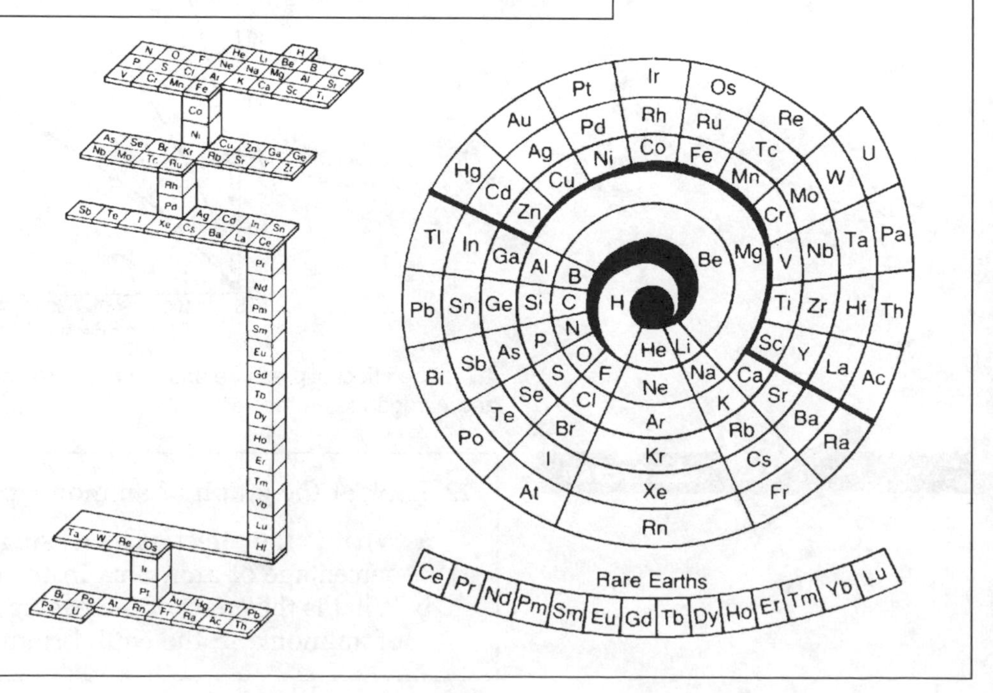

5.2 and 5.3 Two unusual forms of the Periodic Table.

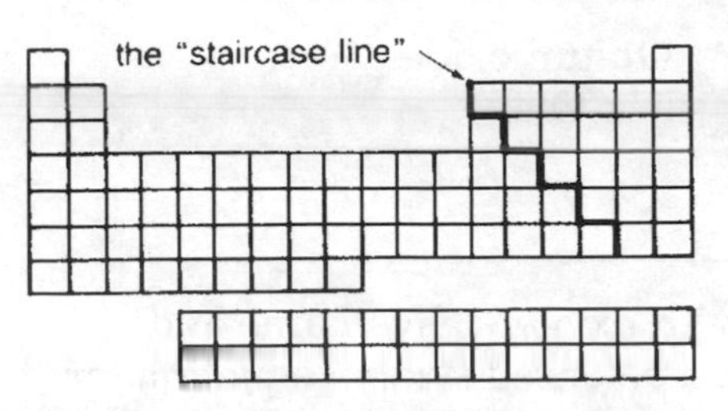

5.4 The 'staircase line' divides metals (on the left) from non-metals (on the right)

Using the Periodic Table

The Periodic Table can tell us a great deal of information about the elements once we know how to use it.

Metals and non-metals

The 'staircase line' divides the metals (on the left) from non-metals (on the right). See Fig 5.4.

What properties would you expect a) the metals and b) the non-metals to have? Look back at chapter 2 if you are not sure.

Some elements near the line appear to have a combination of metallic and non-metallic properties. For example silicon, Si. This element is just on the non-metal side of the line yet it looks quite shiny, rather like a metal, and it conducts electricity to some extent so that it is called a semi-conductor. Its electrical conduction can be altered, which has led to the development of silicon chips, micro-electronics and lap-top computers. Elements like silicon are called **metalloids** or semi-metals.

1 You may never have heard of any of the following elements. Pick from them two which are definitely metals, two non-metals and one which is probably semi-metal:

Ge, Tl, Xe, Sr, F

Groups

These are vertical columns in the Periodic Table. They are usually numbered from left to right missing out the central block of **transition elements**. If you look at the electron arrangements of all the elements within a particular group it is obvious that what they have in common is the same number of outer electrons.

Group I	Group II	Group III	Group 0
			He 2
Li 2,1	Be 2,2	B 2,3	Ne 2,8
Na 2,8,1	Mg 2,8,2	Al 2,8,3	Ar 2,8,8
K 2,8,8,1	Ca 2,8,8,2		

This means that elements in the same group will react in similar ways. They behave like a chemical family.

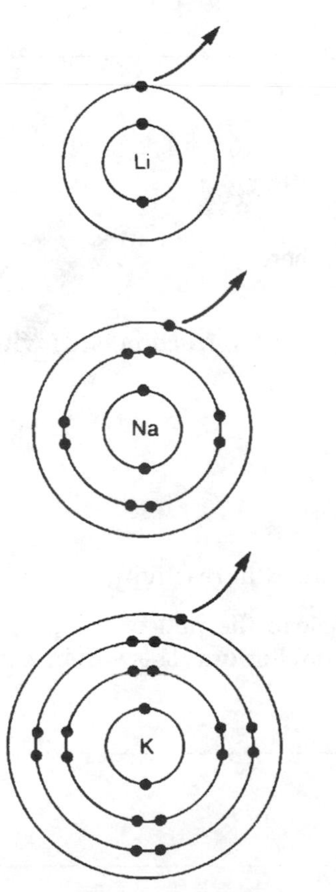

5.5 Group I atoms all lose the outer electron when they react with other atoms.

Group I the alkali metals

Each element reacts by losing its outermost electron, forming a 1+ **ion**. Losing this outer electron takes relatively little energy (the amount is called the **ionisation energy**) and so this group of elements is very reactive.

The elements react with oxygen at room temperature and therefore have to be stored out of contact with the air.

lithium + oxygen → lithium oxide

$4\,Li(s) + O_2(g) \rightarrow 2\,Li_2O(s)$

2 Predict the reaction of sodium with oxygen. Give name and formula of the product and write a balanced symbol equation for the reaction. Remember members of the same group react similarly.

The Group I elements also react with water, producing hydrogen and forming alkaline solutions of their hydroxides. This is the reason for the group's name.

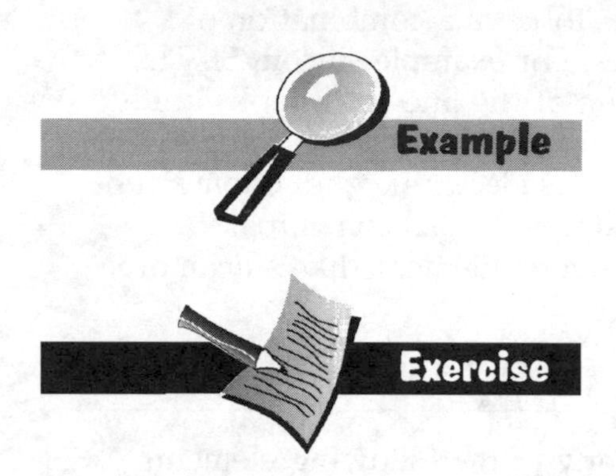

lithium + water → lithium hydroxide + hydrogen

$2\,Li(s) + 2\,H_2O(l) \rightarrow 2\,LiOH(aq) + H_2(g)$

Note: lithium hydroxide is the alkali.

3 Write word and symbol equations for the reactions of a sodium and b potassium with water.

In all its reactions lithium is the least reactive of the alkali metals, sodium is more reactive, and potassium is even more so. We call this a **reactivity trend**. Trends allow us to make predictions

Reactions of all the alkali metals with water

Lithium

The metal floats on a bed of steam produced by the heat of the reaction.

Sodium

The reaction produces so much heat that the sodium melts and forms a molten sphere.

Potassium

Not only does the potassium melt but it ignites the hydrogen which is produced and the reaction ends with a small explosion.

The reactivity trend is clear.

least reactive → most reactive

Li Na K

The atomic structures of lithium, sodium and potassium can help to explain the trend in reactivity.

Each atom reacts by losing one electron, but the electron in lithium is the closest to the pull of the positive nucleus. This makes it easier to detach the outermost electron from sodium than from lithium. This will be easier still in the case of potassium.

4 Use the information above to predict whether potassium or caesium is the more reactive. Explain your reasoning.

As well as trends in chemical properties, Group I also has trends in physical properties. Some of these are shown in Table 5.1.

Table 5.1 Properties of the alkali metals

Element	Atomic number	Electron arrangement	Density/ g/cm^3	Melting point/ °C	Boiling point/ °C
Li	3	2, 1	0.53	181	1342
Na	11	2, 8, 1	0.97	98	883
K	19	2, 8, 8, 1	0.86	63	760
Rb	37	2, 8, 18, 8, 1	1.53	39	686
Cs	55	2, 8, 18, 18, 8, 1	1.88	29	669
Fr	87	2, 8, 18, 18, 32, 8, 1	Can you predict?		

5 The properties of francium are not accurately known because it is so rare. Try to predict its density, melting point and boiling point from the trends in Table 5.1. Hint: one way to approach this is to plot graphs of property against atomic number.

Group II the alkaline earth metals

These metals react by losing two electrons from their outer shells, forming 2+ **ions**. This requires more energy than just losing one electron so these elements are less reactive than the corresponding elements in Group I (but more reactive than those in Group III).

Their chemical properties are broadly similar to those of the Group I elements.

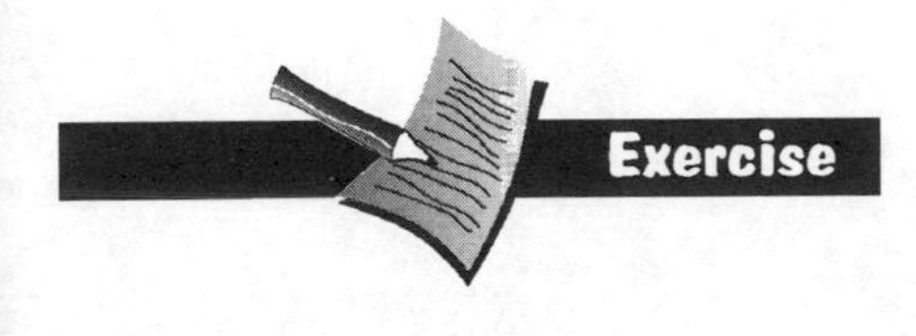

6 What do you expect the reactivity trend to be as you go down Group II from beryllium to radium? Explain your answer.

Group VII the halogens

The halogens are a group of non-metals. The elements are similar to one another in several ways:

- they all exist as diatomic molecules, F_2, Cl_2, etc.
- they are all bleaches and have a 'bleachy' smell.
- they form compounds with similar formulae.

Again there are trends in their properties, some of which are shown in Table 5.2.

Table 5.2 Properties of the halogens

Element	Atomic number	Electron arrangement	Appearance	Melting point/ °C	Boiling point/ °C
F	9	2, 7	Colourless gas	– 220	– 188
Cl	17	2, 8, 7	Greenish gas	– 101	– 35
Br	35	2, 8, 18, 7	Brown liquid	– 7	59
I	53	2, 8, 18, 18, 7	Black solid	114	184
At	85	2, 8, 18, 32, 18, 7	Can you predict?		

7 Like francium, astatine is a rare and radioactive element whose properties are not accurately known. Use the trends in Table 5.2 to help you predict its missing properties. Hint: one method is to plot a graph of the property against atomic number.

Reactions of the halogens

This time the chemical reactivity is based on getting an extra electron to fill the single gap in the outer shell.

F_2	+	$2e^-$	→	$2F^-$
fluorine molecule				fluoride ions

The reactivity trend in the halogens is as follows.

least reactive → most reactive
At F

This is because the pull of the nucleus is what attracts and holds the extra electron. So the smallest atom, fluorine, has the greatest pull and is the most reactive. Astatine, the largest atom, has the least pull and is least reactive.

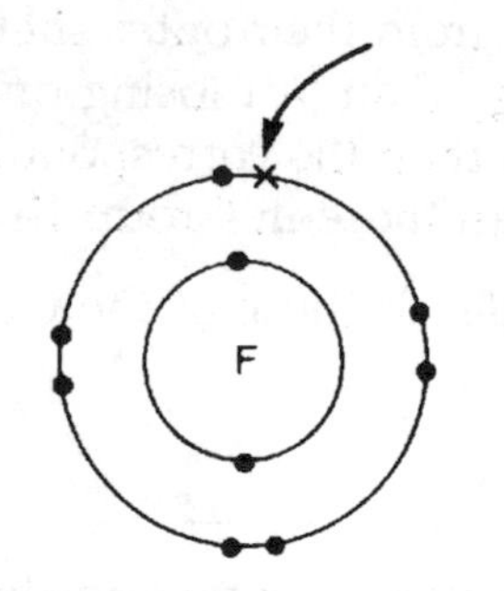

5.6 Group VII atoms gain an electron when they react.

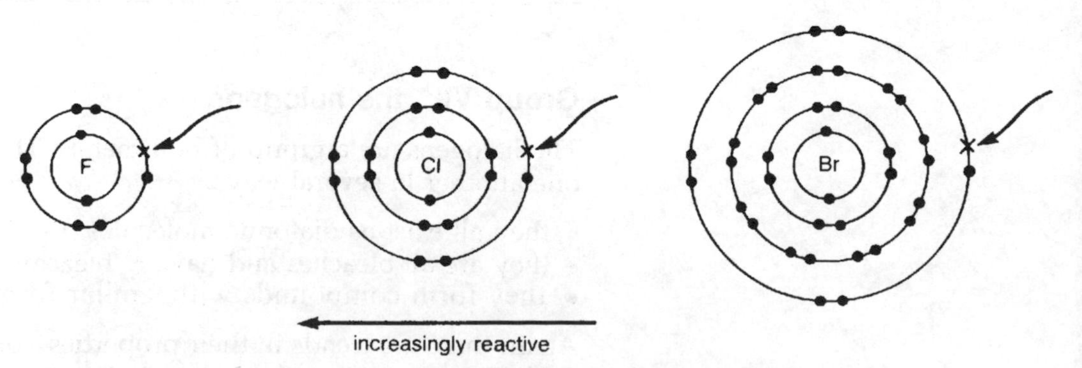

5.7 The reactivity of the elements increases as we go up the group.

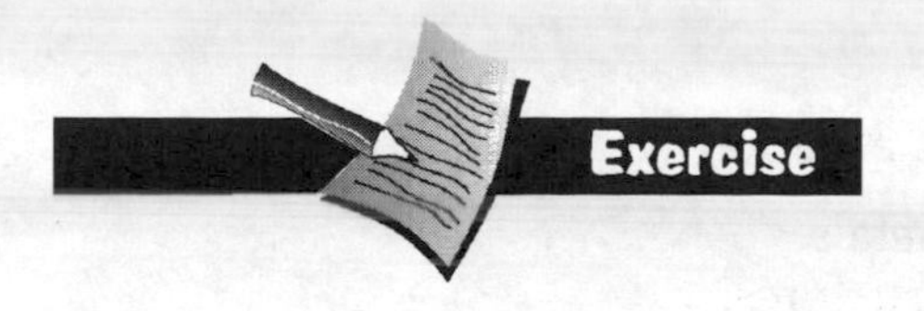

8 Elements in Group VI have to gain two electrons to form a full outer shell. Would you expect Group VI elements to be more or less reactive than the corresponding halogen?

9 The element hydrogen is sometimes placed in Group I but some versions of the Periodic Table place it in Group VII. Use its electron arrangement to explain why both are sensible

The halogens are also able to fill the single gap in their outermost shell by bonding covalently. Group VII atoms can use their unpaired electron to form a single bond with another non-metal atom (see Chapter 2).

Group 0 the inert gases

This group is numbered 0, rather than VIII, because the whole group of elements was discovered after Mendeleev's first version of the Periodic Table and so it was 'tacked on' later. The whole group consists of very unreactive gases. Their unreactivity is due to the fact that they all have complete outer electron shells. They do in fact form a few compounds but the first was discovered only as recently as 1962.

Periods

These are horizontal rows in the Periodic Table.

Period 1 contains hydrogen and helium
Period 2 contains lithium, beryllium, boron, carbon, nitrogen, oxygen, fluorine and neon
and so on

As we go from left to right across a period, the outer shell of electrons is filling up, see Chapter 2.

Periods have patterns too, see Table 5.3.

Table 5.3 Trends across Period 3

Group	I	II	III	IV	V	VI	VII	O
Element	Na	Mg	Al	Si	P	S	Cl	Ar
	Metals			Semi-metal	Non-metals			(Inert gas)
	less reactive →				← less reactive			
Structure of element	Giant metallic			Giant covalent	Molecular			Atomic
Ion	Na^+	Mg^{2+}	Al^{3+}	None	P^{3-}	S^{2-}	Cl^-	None
Oxide	Na_2O	MgO	Al_2O_3	SiO_2	P_4O_{10} (P_4O_6)	SO_2 (SO_3)	Cl_2O (Cl_2O_7 etc.)	None
	Strongly basic		Some acidic and some basic properties	Acidic				
Structure of oxide	Giant ionic			Giant covalent	Molecular			None

As we go from left to right across a period we move from

- metals to semi-metals to non-metals.
- giant structures to molecular structures.
- basic oxides to acidic oxides.
- elements which form positive ions to elements which form negative ions.

The transition metals

The transition elements form a block, not a group or a period. In it are contained most of the everyday metals such as iron and copper. These elements have typical metallic properties (see Chapter 2). They are much less reactive than the metals in Groups I and II which makes them more useful – you could hardly make a car out of sodium! They often form coloured compounds – a property which is caused by their incomplete inner electron shells (these are the 'reserve shells' mentioned in Chapter 2).

Transition metals and their compounds are often used as catalysts. For example, iron is used as a catalyst in the manufacture of ammonia and vanadium oxide is used when making sulphuric acid.

Chapter 6 Chemical calculations

STARTING POINTS

- Do you understand the following terms: **element, compound, atom, molecule, ion, giant structure**? Write a sentence or two (no more) to explain what you understand by each term. Then check the glossary at the back of the book.

The masses of atoms

One of the most important things that chemists did in the last century (and it was a painstaking task) was to find out the *relative* masses of the atoms of the elements. That is, they did not know the actual mass of any atom, but they found out that an oxygen atom was 16 times the mass of a hydrogen atom, while a sulphur atom was twice the mass of an oxygen atom and so on. In this way they formed a scale of mass, with hydrogen as the lightest atom and all the other elements on the scale compared with hydrogen. These masses of the atoms of elements are called their **relative atomic masses** and the symbol for them is A_r. We can look them up in the Periodic Table as shown below.

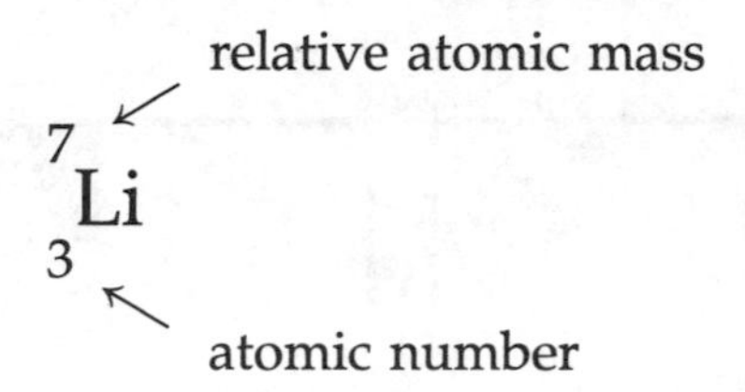

The Periodic Table in this book has most relative atomic masses rounded to the nearest whole number. Other versions of the Periodic Table may have the relative atomic mass and the atomic number reversed but this should not put you off: The relative atomic mass is always the larger.

1 Use the Periodic Table to look up the relative atomic masses of the following atoms:

a sodium, Na c lead, Pb
b oxygen, O d chlorine, Cl

2 Imagine you had an atomic see-saw. A magnesium atom is placed on one side. What combinations of other atoms could be placed on the other side to balance it? For example two carbon atoms would work. There are lots of combinations. Give at least six.

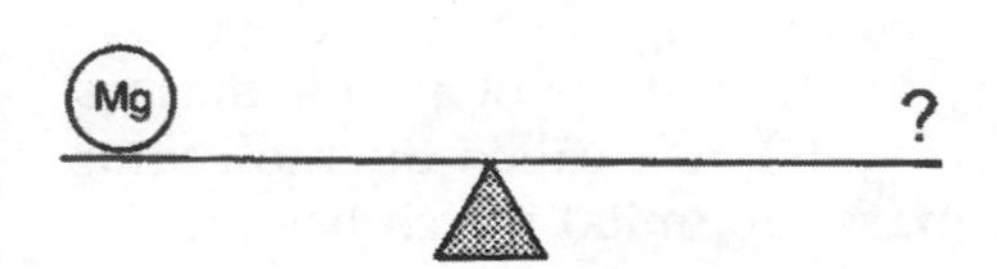

6.1 What combinations of atoms would balance this 'atomic see-saw'?

Atomic masses in u

It is now possible to weigh (measure the masses of) individual atoms using an instrument called a mass spectrometer. The mass of the lightest atom, hydrogen, is 1.6×10^{-27} kg. For simplicity chemists call this mass 1 **atomic mass unit** (symbol u). So the relative atomic mass, A_r, of a hydrogen atom is 1 and its mass is 1 u. Either way the *numbers* are the same. We can, of course, do this for all the elements. A helium atom has four times the mass of a hydrogen atom, so helium's relative atomic mass is 4 and a helium atom has a mass of 4 u. A_r for lithium is 7, and a lithium atom has a mass of 7 u.

The mass spectrometer

How can we weigh atoms? While the details of the mass spectrometer are complicated, the principles are fairly straightforward. Atoms are first bombarded with a beam of electrons which knocks out an electron from each atom to make it into a positive ion. These positive ions are then attracted by a negatively charged plate. This accelerates them to a high speed. Exactly how fast depends on the mass of the atom. The heavier the atom the slower it goes. The speeding ions pass through a hole in the plate and into a magnetic field. A magnetic force then deflects the ions from their straight line path. Again the amount of deflection depends on the mass of the atom. The ions then hit a detector. We can work back from the position where the ions hit the detector to find the mass of the original atom.

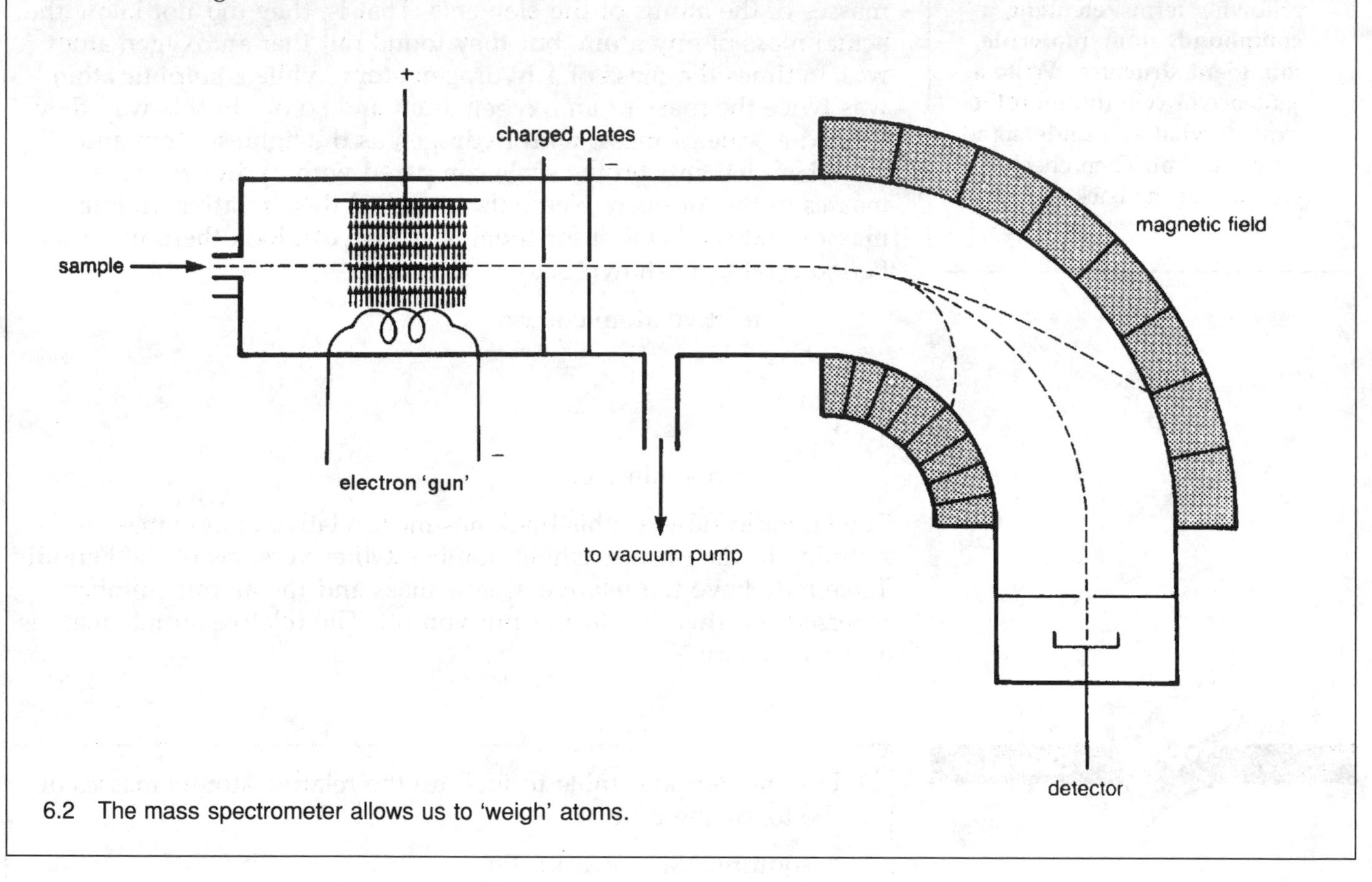

6.2 The mass spectrometer allows us to 'weigh' atoms.

Masses of molecules

The mass of a molecule is found by adding together the masses of the atoms it contains. We can either work in terms of **atomic masses** (in u) or **relative atomic masses** (no units). It is called the **molecular mass** (in u), or **relative molecular mass** (no units) which has the symbol M_r.

A hydrogen fluoride, HF

Atomic masses H = 1 u, F = 19 u
Molecular mass 1 u + 19 u = 20 u
Or relative molecular mass, $M_r = 20$

B water, H_2O

Atomic masses H = 1 u, O = 16 u
Molecular mass = $(2 \times 1) + (1 \times 16) = 18$ u
Or relative molecular mass, $M_r = 18$

C urea $CO(NH_2)_2$

Atomic masses C = 12 u, H = 1 u, N = 14 u, O = 16 u
Molecular mass = 12 + 16 + (14 + [2 × 1]) × 2 = 60 u
Or relative molecular mass, $M_r = 60$

For compounds which do not exist as molecules such as sodium chloride (NaCl) which has a giant structure of ions (see Chapter 2) the term **molecular mass** or **relative molecular mass** is still used even though we are not talking about molecules. If we use the same term it makes everything simpler.

NaCl 23 + 35.5 = 58.5
$CaCO_3$ 40 + 12 + (3 × 16) = 100
$Cu(NO_3)_2$ 64 + (14 + [3 × 16]) × 2 = 188

3 Work out (using relative atomic masses in the Periodic Table) the relative molecular mass of:

a ammonia, NH_3
b calcium hydroxide, $Ca(OH)_2$
c oxygen molecule, O_2
d ethanol, C_2H_6O

Using chemical equations

We can use a **balanced equation** (see Chapter 4) to work out how much (by mass) of different substances react together. For example methane (natural gas) burns according to the following equation:

CH_4	+	$2O_2$	→	$2H_2O$	+	CO_2
1 particle		2 particles		2 particles		1 particle
16 u		2 × 32 u		2 × 18 u		44 u
16 u		64 u		36 u		44 u

Notice that the total mass on the left of the equation is the same as the total mass on the right (80 u). It does not matter what mass units we use. So 16 u of methane will react with 64 u of oxygen to give 36 u of water and 44 u of carbon dioxide; 16 tonnes of methane will react with 64 tonnes of oxygen to give 36 tonnes of water and 44 tonnes of carbon dioxide and so on.

4 Work out the reacting masses in grams for:

a $2H_2 + O_2 \rightarrow 2H_2O$
b $H_2 + Cl_2 \rightarrow 2HCl$

The mole

The examples above show us that substances react together in rather complicated proportions by mass – in the methane example above, 16 : 64 : 36 : 44. To make things simpler, we use a quantity called a **mole** which is the atomic or molecular mass of a substance in grams.

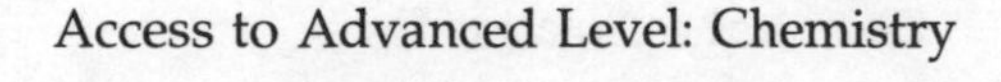

Using the idea that a mole of substance is its molecular mass in grams, find the mass of a mole of water, H_2O.

Molecular mass = 18 u, so a mole of water has a mass of 18 g.

5 Find the mass of a mole of:

a Sn atoms
b NH_3 molecules
c CO_2 molecules
d NaCl

How many molecules are there in 4 g of hydrogen, H_2?

One mole has a mass of 2 g, so number of moles = 4/2 = 2 moles.

In general, the number of moles in a given mass of substance = mass in g/mass of one mole in g.

6 How many moles are there in

a 40 g of calcium, Ca?
b 980 g of sulphuric acid, H_2SO_4?
c 22 g of carbon dioxide, CO_2?
d 3.2 g of oxygen, O_2?

Moles and equations

In the burning of methane example we have 1 mole of methane reacting with 2 moles of oxygen molecules to produce 2 moles of water and 1 mole of carbon dioxide – much simpler than working in grams or u.

CH_4	+	$2\,O_2$	→	$2\,H_2O$	+	CO_2
1 particle		2 particles		2 particles		1 particle
16 u		64 u		36 u		44 u
16 g		64 g		36 g		44 g
1 mole		2 moles		2 moles		1 mole

7 For the following equations, write below each substance the quantity which will react in (i) grams and (ii) moles.

a $CuCO_3$	→	CuO	+	CO_2		
b Mg	+	2 HCl	→	$MgCl_2$	+	H_2
c HCl	+	NaOH	→	NaCl	+	H_2O

Substances react in simple whole number proportions of moles.

This is not a coincidence: a mole of *any* substance contains the same number of particles (atoms, molecules, ions).

If we weigh out a mole of carbon (12 g) and a mole of lead (207 g) those masses have the same number of particles in them. This is because the ratio of the masses of the moles is the same as the ratio of the masses of the atoms.

6.3 These piles of 1 mole of carbon (left) and 1 mole of lead (right) contain the same number of atoms.

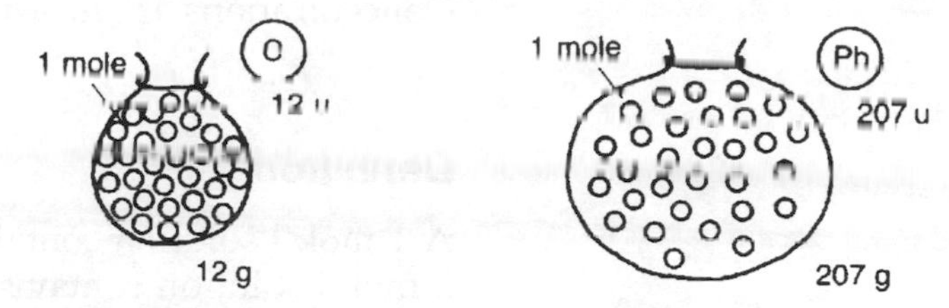

6.4 Each bag contains the same number, 6×10^{23}, of atoms and we call this amount a mole. This gives us a way of counting by weighing.

Here is another way of showing the same thing:
1 u is 1.6×10^{-24} g, so 1 mole of H atoms (1 g) contains $1/1.6 \times 10^{-24}$ atoms. This is 6×10^{23} atoms.

In the same way, 1 mole of He atoms (4 g) contains $4/6.4 \times 10^{-24}$ atoms. This is 6×10^{23} atoms again.

- Note An atom of hydrogen has a mass of $1\ u = 1.6 \times 10^{-24}$ g.
  An atom of helium has a mass of $4\ u = 4 \times 1.6 \times 10^{-24}$ g $= 6.4 \times 10^{-24}$ g.

So 1 mole of any substance contains the same number (6×10^{23}) particles.

Since atoms and molecules react in simple whole number proportions: 1:1, 2:1, 3:1, 3:2, etc., substances react in simple whole number proportions of moles. This is why the idea of the mole is so useful to chemists.

The formal definition of the mole is the quantity of substance which contains the same number of particles as there are atoms in 1 g of hydrogen (or strictly, 12 g of ^{12}C – see Chapter 2). However in practical terms you will find it easier to think of a mole as the atomic or molecular mass in grams or 6×10^{23} particles. The only acceptable abbreviation for mole is mol.

Using the idea of moles

Finding formulae

18 g of water contains 2 g of hydrogen combined with 16 g of oxygen, A_r O = 16, A_r H = 1. What is its formula?

2 g of hydrogen is 2/1 = 2 mol
16 g of oxygen is 16/16 = 1 mol

So 2 mol of hydrogen combines with 1 mol of oxygen. Since a mole of all substances contains the same number of particles, 1 *atom* of oxygen combine with 2 *atoms* of hydrogen: the formula must be H_2O.

8 What is the formula of each of the following compounds?

a 196 g of sulphuric acid contains 4 g of hydrogen, 64 g of sulphur and 128 g of oxygen.
b 12.4 g sodium oxide contains 9.2 g of sodium and 3.2 g of oxygen.
c 17 g of ammonia contains 14 g of nitrogen and 3 g of hydrogen.
d 6.3 g of nitric acid contains 0.1 g of hydrogen, 1.4 g of nitrogen and 4.8 g of oxygen.

Concentration

We use the mole to define **concentration** – how much of a substance (called the solute) is dissolved in water (or other solvent) when we make a solution.

Concentrations are measured in moles per litre (mol/l)

Definition

A 1 mol/l solution contains 1 mole of solute per litre (1000 cm^3). So a 2 mol/l solution contains 2 moles of solute per litre and so on.

9 What is the concentration in mol/l of

a 2 moles of H_2SO_4 in 1000 cm^3 of solution?
b 2 moles of H_2SO_4 in 2000 cm^3 of solution?
c 0.5 moles of H_2SO_4 in 100 cm^3 of solution?

10 How would you make a

a 0.1 mole per litre solution of copper sulphate (relative molecular mass 160)?
b 1 mole per litre solution starting with 16 g of copper sulphate?

We may often need to know how many moles of solute are dissolved in a certain volume of solution.

For example, how many moles of silver nitrate are dissolved in 25 cm^3 of 0.1 mol/l solution?

This can be worked out as follows:

1000 cm^3 of solution contains 0.1 mol of solute

1 cm^3 of solution contains 0.1/1000 mol of solute

25 cm^3 of solution contains (0.1/1000) × 25 mol of solute

This is 2.5×10^{-3} mol.

We can use the formula:

number of moles of solute $= M \times V/1000$,
where M is the concentration of the solution in mol/l and V the volume of the solution in cm^3.

Check that you get the same answer as above by using the formula.

Notice that the calculation does not have anything to do with the *nature* of the solute – only its concentration – so that the answer would be the same whatever the solute.

11 How many moles of solute are there in

a 10 cm^3 of 1 mol/l solution?
b 50 cm^3 of 2 mol/l solution?
c 20 cm^3 of 0.01 mol/l solution?

Making solutions

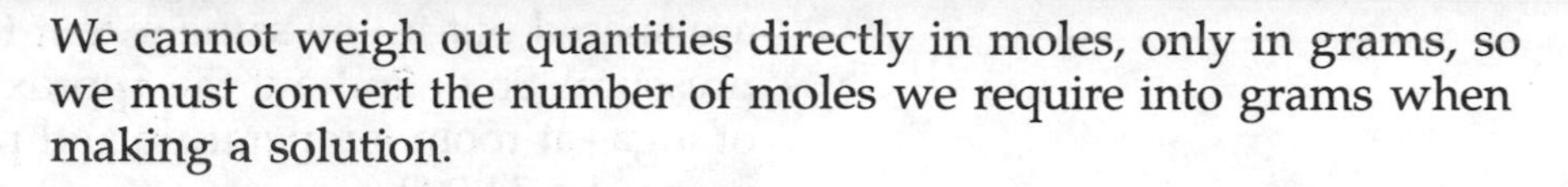

We cannot weigh out quantities directly in moles, only in grams, so we must convert the number of moles we require into grams when making a solution.

Example

How do we make 500 cm^3 of a 0.1 mol/l solution of sodium hydroxide, NaOH?

$$\text{No. of moles} = M \times V/1000$$
$$= 0.1 \times 500/1000$$
$$= 0.05 \text{ mol}$$

The relative molecular mass of sodium hydroxide (NaOH) is $23+16+1=40$, so 1 mole of sodium hydroxide has a mass of 40 g. So we must dissolve $0.05 \times 40 = 2$ g of solid sodium hydroxide and add water until the volume is 500 cm^3. Note that this is not the same as dissolving 2 g of solid in 500 cm^3 of water – this would give a total volume of more than 500 cm^3.

When making a solution, the solute we are using *does* affect the mass of substance we must weigh out. What mass of *potassium* hydroxide would we need to weigh out to make 500 cm^3 of 0.1 mol/l solution?

12 Describe how you would make the following.
- a 1 dm^3 of 2 mol/l sodium chloride solution (NaCl).
- b 2 dm^3 of 0.01 mol/l of silver nitrate solution ($AgNO_3$).
- c 250 cm^3 of 0.5 mol/l of sodium thiosulphate solution ($Na_2S_2O_3$).
- d 100 cm^3 of 1 mol/l iodine solution (I_2).

Using concentrations

Finding the proportions in which substances react.

100 cm^3 of 1 mol/l hydrochloric acid is completely neutralised by 50 cm^2 of 2 mol/l sodium hydroxide. In what proportion do the two react?

Using the formula no. of moles $= M \times V/1000$
No. moles of hydrochloric acid $= 1 \times 100/1000 = 0.1$ mol
No. moles of sodium hydroxide $= 2 \times 50/1000 = 0.1$ mol
So the substances react 1 : 1.

13 a 100 cm^3 1 mol/l sulphuric acid is just neutralised by 50 cm^3 of 4 mol/l of potassium hydroxide. In what proportion do they react?
b 20 cm^3 of 0.1 mol/l nitric acid is just neutralised by 10 cm^3 of 0.2 mol/l sodium hydroxide. In what proportions to they react?

Moles of gases

It is not easy to measure the mass of a gas but we can measure gas volumes easily. The volume of a mole of *any* gas is about 24 000 cm^3 at normal temperatures and pressures whatever the relative molecular mass. This may seem strange but it is because the particles in all gases are spaced out to the same extent (at the same temperature and pressure). So to find out the approximate number of moles in a sample of a gas at room temperature and pressure we must divide its volume in cm^3 by 24 000.

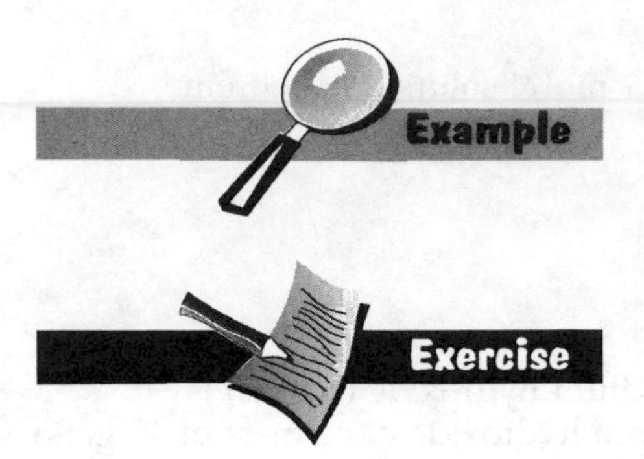

A reaction produces 120 cm^3 of hydrogen gas. How many moles of gas is this?

Moles of gas = 120/24 000 = 0.005 mol

14 How many moles are there in the following volumes of oxygen (O_2) at room temperature and pressure?

a 240 000 cm^3 b 120 cm^3 c 48 cm^3

d What difference (if any) would it make if the gas were neon?

Moles of liquids

Liquids can be weighed out in the same way as solids but it is often more convenient to measure them out by volume, using a pipette or measuring cylinder. In this case we need to use the **density** of the liquid to calculate the volume we need. The density of a substance is the mass in grams of 1 cm^3 (or the mass in kg of 1 m^3).

$$\text{density} = \frac{\text{mass}}{\text{volume}}$$

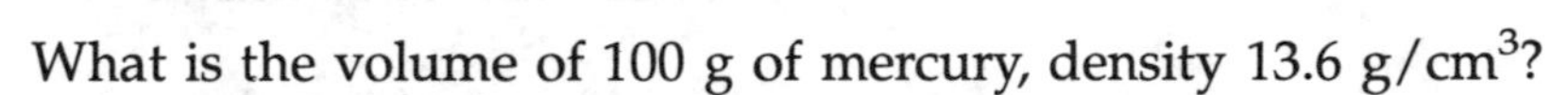

What is the volume of 100 g of mercury, density 13.6 g/cm^3?

density = mass/volume

volume = mass/density

= 100/13.6

volume = 7.35 cm^3

We need to measure out 0.1 of a mole of ethanol whose density is 0.8 g/cm^3. What volume do we need? The relative molecular mass of ethanol (C_2H_6O) is 46. So the mass of 1 mole is 46 g and that of 0.1 mole is 4.6 g.

density = mass/volume

So volume = mass/density

= 4.6/0.8

= 5.75 cm^3

15 Work out the volume of the following.

a 250 g of trichloroethane, density 1.3 g/cm^3

b 1000 g of water, density 1 g/cm^3

c 2 kg of ethanol, density 800 kg/m^3

d 2 moles of trichloroethane, $C_2H_3Cl_3$

e 0.5 mole of water, H_2O

f 2 moles of ethanol, C_2H_6O, density 0.8 g/cm^3

How to measure out quantities in moles – Summary

1 Solids
Use no. of moles = mass in grams/relative molecular mass

2 Liquids
Use no. of moles = mass in grams/relative molecular mass and density = mass/volume

3 Solutions
Use no. of moles of solute = $M \times V/1000$

4 Gases
At room temperature and pressure use the approximate expression no. of moles = volume in $cm^3/24\ 000$

What use are chemical calculations? (1)

Copper ores
Copper can be extracted from two ores – copper carbonate, $CuCO_3$, (malachite) or copper iron sulphide, $CuFeS_2$, (copper pyrites). Which contains the greater percentage of copper?

Malachite
The relative molecular mass of $CuCO_3$ is $64 + 12 + (3 \times 16) = 124$.
So 1 mole has a mass of 124 g.
Of this, 64 g is copper. So the % of copper is $64 \times 100/124 = 51.6\%$.

Copper pyrites
The relative molecular mass of $CuFeS_2$ is $64 + 56 + (2 \times 32) = 184$.
So 1 mole has a mass of 184 g.
Of this, 64 g is copper. So the % of copper is $64 \times 100/184 = 34.8\%$.
So malachite is the better ore in terms of % copper. However, copper pyrites also contains some iron. Calculate the % of iron.
What other factor would you need to take into account when deciding which ore to buy for processing?

What use are chemical calculations? (2)

Analysing antacids

Antacid tablets can be taken to cure acid indigestion. Some types contain magnesium hydroxide, an alkali, which reacts with hydrochloric acid, the acid present in our stomachs, according to the equation below.

$$Mg(OH)_2 + 2HCl \rightarrow MgCl_2 + 2H_2O$$

One brand of antacid is claimed to contain 0.3 g of magnesium hydroxide per tablet. One tablet sampled from the production line was found to react with 10 cm^3 of 1 mol/l hydrochloric acid. Is the tablet up to specification?

No. of moles of hydrochloric acid = $M \times V/1000$
= $1 \times 10/1000 = 0.01$ mol

From the equation, each mole of hydrochloric acid reacts with $\frac{1}{2}$ mole of magnesium hydroxide. So the tablet must contain $0.01/2 = 0.005$ mol of magnesium hydroxide.

The relative molecular mass of $Mg(OH)_2$ is $24 + [2 \times (16 + 1)] = 58$.

So 1 mole of magnesium hydroxide has a mass of 58 g.

The tablet contains 0.005 mol so it contains $0.005 \times 58 = 0.29$ g.

The tablet is below specification and should be rejected.

Devise a practical method for carrying out this quality control procedure. Try to write out the method so that it could be followed by one of your classmates on work experience.

Further calculations

Equations where you have to use the idea of the mole to solve the problem are very common at Advanced Level. Go slowly until you have gained confidence. There are short cuts, but it is wise to really understand each step. This set of questions will also test your knowledge of formulae and of acid reactions.

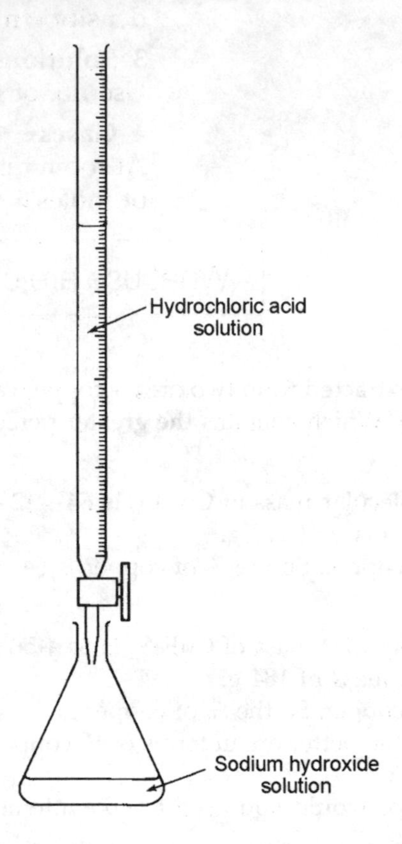

6.5 Neutralising sodium hydroxide.

Acid reactions

A 25 cm^3 of a solution of sodium hydroxide is neutralised by 20 cm^3 of 0.1 mol/l hydrochloric acid. What is the concentration, *C*, in mol/l of the sodium hydroxide solution?

First write the balanced symbol equation – it is very important to get this correctly balanced.

	NaOH	+	HCl	→	H_2O	+	NaCl
Put in the number of moles	1		1		1		1
	0.002		*0.002*		*0.002*		*0.002*

This tells you that for every mole of sodium hydroxide used, the *same* number of moles of hydrochloric acid is used.

Work out the number of moles of hydrochloric acid:

$$\text{No. of moles of HCl} = \frac{M \times V}{1000} = \frac{0.1 \times 20}{1000} = 0.002$$

The equation tells us that the number of moles of NaOH must also equal 0.002.

(It is a good idea at this stage to write the number of reacting moles under the equation.)

We also know:

$$\text{No. of moles of NaOH} = \frac{M \times V}{1000} = \frac{C \times 25}{1000}$$

$$\text{So } \frac{C \times 25}{1000} = 0.002$$

$$\text{From this } C = \frac{0.002 \times 1000}{25}$$

$$\text{and } C = 0.08 \text{ mol/l}$$

it is a useful trick here to do a bit of mental calculation. Does this answer make sense? Well, we started with 20 cm^3 of 0.1 mol/l acid and we know that we are going to need the same number of moles of sodium hydroxide. We used 25 cm^3 of this (slightly more than the acid), so it must have been a bit less concentrated.

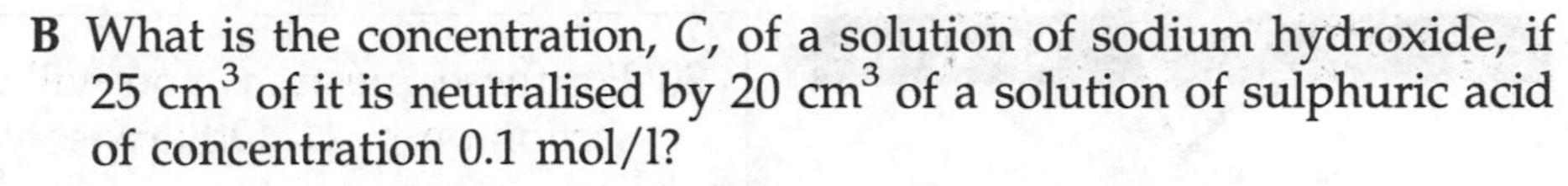

Example

B What is the concentration, *C*, of a solution of sodium hydroxide, if 25 cm^3 of it is neutralised by 20 cm^3 of a solution of sulphuric acid of concentration 0.1 mol/l?

Write the balanced symbol equation:

	$2NaOH$ +	H_2SO_4 →	Na_2SO_4 +	$2H_2O$
Put in the number of moles	2	1	1	2
	0.004	*0.002*	*0.002*	*0.004*

(Notice that we have twice as many moles of sodium hydroxide as sulphuric acid.)

$$\text{Work out the no. of moles of } H_2SO_4 = \frac{M \times V}{1000} = \frac{20 \times 0.1}{1000} = 0.002$$

Add this information to the equation and notice that this time we have 0.0004 moles of NaOH for every 0.002 moles of H_2SO_4

$$\text{No. of moles of NaOH} = 0.004 = \frac{25 \times C}{1000}$$

And from this $C = 0.16$ mol/l

Exercise

16 What is the concentration of a solution of hydrochloric acid if 25 cm^3 of it were neutralised by 10 cm^3 of a solution of sodium hydroxide of concentration 0.5 mol/l?

17 What is the concentration of a solution of sulphuric acid if 20 cm^3 of it were neutralised by 30 cm^3 of a solution of sodium hydroxide of concentration 0.2 mol/l?

18 What volumes of the following solutions will react with 25 cm^3 of 0.2 mol/l sulphuric acid?

a 0.25 mol/l sodium carbonate solution.
Hint – use the same method but this time you are trying to find the volume, *V*, of the solution rather than its concentration.

b 0.02 mol/l potassium hydroxide solution.

19 What mass of the following will react completely with 10 cm^3 of 0.5 mol/l hydrochloric acid?

a magnesium
b zinc
c copper oxide
d potassium carbonate

Check the formulae in Chapter 3.
The first one is done for you as a worked example.

Write the equation with the number of moles of each species:

Mg	+	$2\,HCl$	→	$MgCl_2$	+	H_2
1		2		1		1
0.0025		*0.005*		*0.0025*		*0.0025* ←

Work out the no. of moles of acid $= M \times V = \frac{0.5 \times 10}{1000} = 0.005$

Add the number of reacting moles.

Now you must work out the mass of 0.0025 moles of Mg.

The A_r of Mg is 24 so the mass of 1 mole is 24 g.

So the mass of 0.0025 moles of Mg is 24×0.0025 g $= 0.06$ g.

20 For these questions you will need to refer to the Moles of gases section on p. 41. Otherwise you tackle them in the same way.

a What volume of carbon dioxide is released when 1 g of calcium carbonate reacts with excess nitric acid at room temperature and pressure?

b What volume of hydrogen is released when excess magnesium reacts with 10 cm^3 of 2 mol/l sulphuric acid at room temperature and pressure?

'Excess' means that there is more than enough of one reactant to react with all of the other.

Chapter 7 Organic chemistry

STARTING POINTS

- Do you understand the following terms: **covalent bond, alkane, alkene**? Write a sentence or two (no more) to explain what you understand by each term. Then check the glossary at the back of the book.

Organic chemistry is the chemistry of compounds based on chains and rings of carbon atoms. There are more known compounds of the element carbon than of all the other elements put together – about ten million at the latest count. This huge variety of compounds exists for several reasons. Carbon has four electrons in its outer shell and can therefore form four covalent bonds (Fig 7.1). These bonds can be formed with other carbon atoms or with different elements. This means that carbon atoms can form chains where two of the four bonds are with other carbon atoms. This still leaves two bonds 'spare' so that the chain can have branches. Another possibility is for the ends of the chain to link together forming a ring (Fig 7.2).

Carbon forms strong bonds with other carbon atoms and also with other elements such as hydrogen, oxygen, sulphur, nitrogen and the halogens. This means that carbon-based compounds are stable.

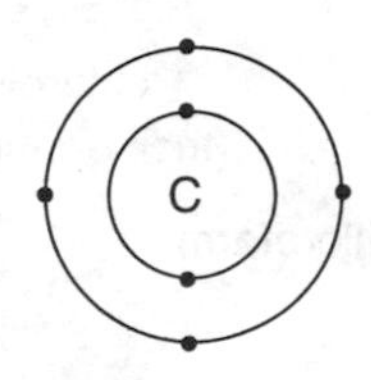

7.1 The electronic structure of carbon. It has four electrons in its outer shell and can therefore form four covalent bonds

```
  H  H  H  H  H        H  H  H  H             H   H
  |  |  |  |  |        |  |  |  |           H  \ /  H
H-C--C--C--C--C-H    H-C--C--C--C-H             C
  |  |  |  |  |        |  |  |  |            \ / \ /
  H  H  H  H  H        H  H  |  H         H-C     C-H
                          H-C-H           H-C     C-H
                            |                \ / \ /
                            H                 C
                                            H / \ H
                                              H   H
```

7.2 Carbon compounds can form chains, chains with branches and rings.

Displayed formulae of organic compounds

The shapes of organic compounds are often very important so we need to use a way of drawing organic molecules which shows this. We often use **displayed** formulae which show covalent bonds as dashes. So

```
               H
               |
methane      H-C-H              CH4
               |
               H

               H  H
               |  |
ethane       H-C--C-H           C2H6
               |  |
               H  H

               H  H  H
               |  |  |
propane      H-C--C--C-H        C3H8
               |  |  |
               H  H  H

               H  H  H  H
               |  |  |  |
butane       H-C--C--C--C-H     C4H10
               |  |  |  |
               H  H  H  H
```

methane CH_4, ethane C_2H_6, propane C_3H_8, butane C_4H_{10}

```
                   H   H   H
                   |   |   |
methylpropane  H — C — C — C — H
                   |   |   |
                   H   |   H
                     H—C—H
                       |
                       H
```

Double bonds are shown as C=C and triple bonds as C≡C.

Even this is only a shorthand; organic compounds are really three-dimensional molecules, as shown in Fig 7.3.

To really get a good impression of the shapes of organic molecules, we need to use models or computer graphics (Fig 7.4).

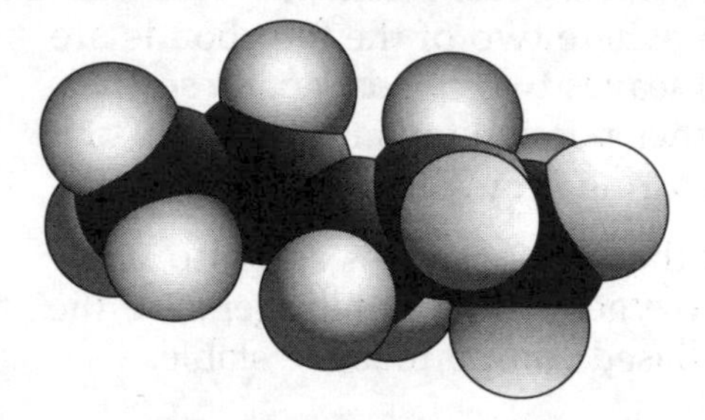

7.3 A three-dimensional representation of pentane (C_5H_{12}).

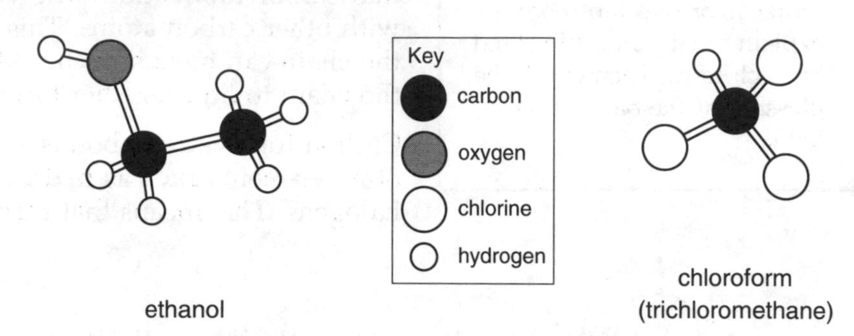

7.4 Computer graphic representation of ethanol and chloroform.

Fig 7.5 shows the shapes of some carbon compounds important in everyday life.

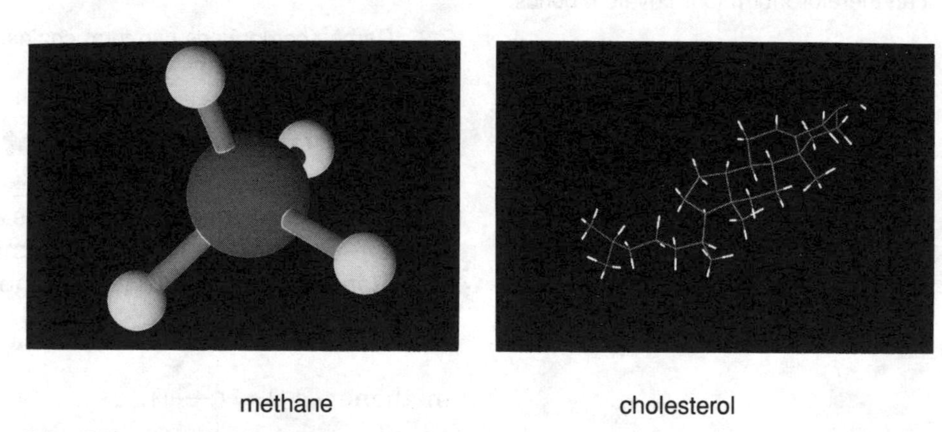

7.5 Different styles for representing the shapes of organic molecules.

Carbon, the element of life

Most of the complex molecules found in living things are based on carbon, so carbon may be called the element of life. Scientists once thought that compounds formed in living things were in some way different from those from non-living sources. This was the origin of the word 'organic' meaning to do with living organisms. Friedrich Wöhler was the first chemist to disprove this idea when he made urea from ammonium thiocyanate. Ammonium thiocyanate had not been made by anything living. Urea is an organic compound produced by living things. The word 'organic' has gradually changed its meaning for chemists so that it is now taken to describe molecules based on carbon.

Naming organic compounds

Because of the variety of carbon compounds, the simple system used for inorganic compounds (copper oxide, lead chloride, etc.) is no use – there would be hundreds of different compounds called carbon hydride, for example. We use a system based on names which are related to the length of the main carbon chain. The stems of these names are

meth-	one carbon
eth-	two carbons
prop-	three carbons
but-	four carbons
pent-	five carbons

After but-, the names are related to those of geometrical figures such as hexagons, etc.

Endings are added to these stems to tell us about the type of compound – the 'functional group'.

Some common ones are:

-ane	C–C and C–H bonds only, an alkane
-ene	one or more C=C bonds, an alkene
-yne	one or more C≡C bonds, an alkyne
-ol	one or more –O–H groups, an alcohol

So ethanol is an alcohol with two carbon atoms and propene the alkene with three carbon atoms.

```
    H   H                  H   H      H
    |   |                  |   |     /
H — C — C — O — H      H — C — C == C
    |   |                  |         \
    H   H                  H          H
```

ethanol   propene

Sometimes more information is needed. For example in propanol, the –OH group could be on the end carbon or the middle one. We use a number, called a **locant**, to tell us where it is. This counts from the end of the chain. So

```
        H   H   H
        |   |   |
H - O - C - C - C - H
        |   |   |
        H   H   H
```

is propan-1-ol and

```
    H   H   H
    |   |   |
H - C - C - C - H
    |   |   |
    H   |   H
        O - H
```

is propan-2-ol.

```
        H   H   H                  H   H   H
        |   |   |                  |   |   |
H - O - C - C - C - H          H - C - C - C - O - H
        |   |   |                  |   |   |
        H   H   H                  H   H   H
```

The two formulae above are the same substance (simply turned round on the paper) so they must have the same name. It is propan-1-ol (not propan-3-ol) as we always use the smallest possible locant.

Branches

Often the carbon chain is branched as in the compound below.

```
    H   H   H   H
    |   |   |   |
H — C — C — C — C — H
    |   |   |   |
    H   |   H   H
      H-C-H
        |
        H
```

To name branched-chain compounds, we pick the longest unbranched chain and this gives us the basic stem of the name. In this case the longest unbranched chain has four carbons so the stem name is but-. Branches are named according to their length and have the ending -yl. Here we have a one-carbon branch so it is called methyl-. There are no double bonds or other functional groups so the compound is called methylbutane. Note that this is written as one word with no hyphen.

Sometimes there might be several branches. If so we list them in alphabetical order – ethyl before methyl, for example, and we may need locants to say where they are. More than one branch of the same length is shown using the prefixes di- (two), tri- (three), tetra- (four).

So

```
  H  H  H       H  H  H  H  H
  |  |  |       |  |  |  |  |
H-C--C--C-------C--C--C--C--C-H
  |  |  |       |  |  |  |  |
  H  H  |       |  H  H  H  H
      H-C-H     |
        |     H-C-H
        H       |
              H-C-H
                |
                H
```

is 4-ethyl-3-methyloctane. Note that we write hyphens between numbers (locants) and letters.

and

```
                 H
                 |
  H  H  H      H-C-H H  H  H  H
  |  |  |        |   |  |  |  |
H-C--C--C--------C---C--C--C--C-H
  |  |  |        |   |  |  |  |
  H  H  |      H-C-H H  H  H  H
      H-C-H      |
        |        H
      H-C-H
        |
        H
```

Ethyl is listed before methyl as the rule is to list groups alphabetically. We place a comma between two successive numbers.

is 3-ethyl-4,4-dimethyloctane.

1 Match the following names with the formulae.

a hexane
b 2-methylpentane
c 3-methylpentane
d 2,2-dimethylpentane

```
  H  H  H  H  H  H
  |  |  |  |  |  |
H-C--C--C--C--C--C-H
  |  |  |  |  |  |
  H  H  H  H  H  H
        (1)
```

```
          II
          |
        H-C-H
  H  H    |  H  H
  |  |    |  |  |
H-C--C----C--C--C-H
  |  |    |  |  |
  H  H    H  H  H
         (2)
```

```
     H
     |
   H-C-H
  H  |  H  H  H
  |  |  |  |  |
H-C--C--C--C--C-H
  |  |  |  |  |
  H  H  H  H  H
        (3)
```

```
            H
            |
          H-C-H
  H  H  H   |  H
  |  |  |   |  |
H-C--C--C---C--C-H
  |  |  |   |  |
  H  H  H   |  H
          H-C-H
            |
            H
           (4)
```

How do we find the formulae of organic compounds?

Organic compounds are mostly made up of carbon, hydrogen, oxygen and nitrogen. One way of finding their formulae is to burn them in a special apparatus so that all the carbon is turned into carbon dioxide, all the hydrogen into water and all the nitrogen becomes uncombined nitrogen. We then measure the amounts of these products, either by weighing them or by more sophisticated instrumental methods.

Here is an example. 2.8 g of compound X burned to form 8.8 g of carbon dioxide (CO_2) and 3.6 g of water (H_2O). What is its formula?

The M_r of CO_2 is 44, so 8.8 g is $8.8/44 = 0.2$ mol.

0.2 mol of CO_2 contains 0.2 mol of carbon atoms as each carbon dioxide molecule contains one atom of carbon.

The M_r of H_2O is 18, so 3.6 g is $3.6/18 = 0.2$ mol.

0.2 mol of H_2O contains 0.4 mol of hydrogen atoms as each water molecule contains two atoms of hydrogen.

So compound X has 0.2 mol of C and 0.4 mol of H. This is twice as much hydrogen as carbon, so the simplest formula is CH_2.

Another example. Compound Y burned to form 8.8 g of carbon dioxide, 2.8 g of nitrogen (N_2) and 9.0 g of water. What is its formula?

The M_r of CO_2 is 44, so 8.8 g is $8.8/44 = 0.2$ mol.

0.2 mol of CO_2 contains 0.2 mol of carbon atoms as each carbon dioxide molecule contains one atom of carbon.

The M_r of N_2 is 28 so 2.8 g is $2.8/28 = 0.1$ mol.

0.1 mol of N_2 contains 0.2 mol of nitrogen atoms as each nitrogen molecule contains two nitrogen atoms.

The M_r of H_2O is 18, so 9.0 g is $9.0/18 = 0.5$ mol.

0.5 mol of H_2O contains 1.0 mol of hydrogen atoms as each water molecule contains two atoms of hydrogen.

So compound Y has 0.2 mol of C, 0.2 mol of N and 1.0 mol of H.

The simplest formula is therefore CNH_5.

2 Try these examples. They are all simple ones where the compound contains only hydrogen and carbon.

a Compound A burns to give 8.8 g of carbon dioxide and 5.4 g of water.
b Compound B burns to give 4.4 g of carbon dioxide and 3.6 g of water.

Reactions of organic compounds

There are millions of organic reactions, but fortunately organic compounds fall into 'families' with similar reactive groups. Compounds in the same chemical family tend to react in the same way, so all alcohols (which have an –OH group) react similarly and all alkenes (which have a C=C) have their own characteristic reactions. So, if you know the reactions of ethanol, you will be able to work out the reactions of propanol, for example.

We can cover only a few organic reactions here. You will learn more as you progress through your Advanced Level course.

Alkanes

Alkanes such as C_2H_6, C_3H_8 and so on have few reactions. This is partly because they have only C–C and C–H bonds which are both strong.

Combustion

Alkanes will burn (combine with oxygen) to give carbon dioxide and water. For example

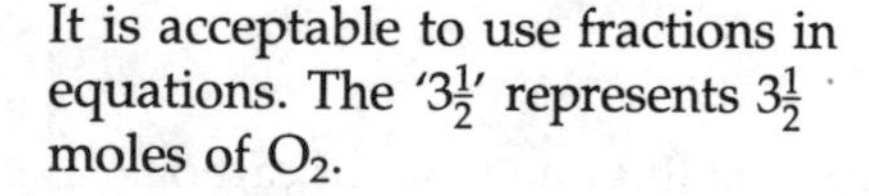

It is acceptable to use fractions in equations. The '$3\frac{1}{2}$' represents $3\frac{1}{2}$ moles of O_2.

ethane	+	oxygen	→	carbon dioxide	+	water
C_2H_6	+	$3\frac{1}{2}O_2$	→	$2CO_2$	+	$3H_2O$

This is an exothermic reaction (it gives out heat). It is of practical importance as many alkanes are used as fuels – methane is natural gas, butane is Calor gas and petrol is a mixture of alkanes of chain length about eight. However it is not a very useful reaction for making new organic compounds.

3 Write balanced equations for the combustion of these other alkanes:

a propane
b butane
c decane.

Cracking

Alkanes can be cracked. This involves using heat and often a catalyst to break the carbon chains to give smaller molecules. When the chain breaks, two new ends are formed. This means that there are not enough hydrogens to make two alkanes so one of the new chains becomes an alkene. Cracking long hydrocarbon chains is useful for two reasons. Firstly, shorter chain alkanes are in greater commercial demand than long chain ones. Secondly, alkenes, being more reactive, are more useful chemically than alkanes.

For example if butane were cracked, possible products are ethane and ethene if the chain breaks in the middle. Another pair of products is methane and propene

```
    H  H ¦ H  H                         H  H           H     H
    |  | ¦ |  |                         |  |            \   /
H — C— C—¦—C— C — H   ——————►     H — C— C — H    +      C=C
    |  | ¦ |  |                         |  |            /   \
    H  H ¦ H  H                         H  H           H     H

    H ¦ H  H  H                           H           H    H   H
    | ¦ |  |  |                           |            \   |   |
H — C—¦—C— C— C — H   ——————►     H — C — H     +      C=C — C — H
    | ¦ |  |  |                           |            /       |
    H ¦ H  H  H                           H           H        H
```

4 Give all the possible products from cracking pentane.

Alkenes

The reactions of alkenes are centred around the double bond. A whole variety of things can add on across the double bond. For example

```
                                          Br  Br
H     H                                   |   |
 \   /                                H — C — C — H
  C=C      +    Br—Br   ——————►           |   |
 /   \                                    H   H
H     H
```

This is often used as a test for an alkene. Bromine is brown and as it is used up in the reaction its colour disappears. So 'decolorising bromine solution' is a test for a compound which has a C=C.

Other things can add on across the double bond, for example hydrogen halides (such as hydrogen chloride, hydrogen bromide or hydrogen iodide)

```
                                          H   Cl
H     H                                   |   |
 \   /                                H — C — C — H
  C=C      +    H—Cl    ——————►           |   |
 /   \                                    H   H
H     H
```

and hydrogen.

```
                                          H   H
H     H                                   |   |
 \   /                                H — C — C — H
  C=C      +    H—H     ——————►           |   |
 /   \                                    H   H
H     H
```

In this case the alkene becomes an alkane. A similar reaction is used, with a nickel catalyst, in the manufacture of margarine to 'harden' oils to make them 'spreadable'.

Not surprisingly, these reactions are called **addition** reactions.

5 Write the equation for water being added onto ethene. Hint, the water adds on as H and OH. Name the product.

Polymerisation

One of the most useful reactions of alkenes is that they can react with themselves to form long chain molecules called polymers. Ethene does this to form poly(ethene) (usually called polythene) as shown.

$$H_2C{=}CH_2 + H_2C{=}CH_2 + H_2C{=}CH_2 \longrightarrow -CH_2-CH_2-CH_2-CH_2-CH_2-CH_2-$$

Substituted alkenes (ones in which one of the hydrogens has been replaced by another atom or group of atoms which we shall call R) also form polymers whose properties depend on R.

R	**Polymer**
$-Cl$	polyvinylchloride (PVC)
$-C_6H_5$	polystyrene
$-CH_3$	polypropylene

Alcohols

Substitution reactions

Alcohols can exchange their –OH group for another group or atom such as a halogen:

$$C_2H_5OH \;+\; HBr \;\rightarrow\; C_2H_5Br \;+\; H_2O$$

This type of reaction is called a **substitution** reaction because one atom or group of atoms has been changed for another.

Oxidation reactions

Another typical reaction of alcohols is that they can be oxidised (by addition of oxygen and removal of hydrogen) to give a new functional group called a carboxylic acid:

$$CH_3-CH_2-O-H \xrightarrow{[+O,\,-2H]} CH_3-C(=O)-O-H$$

This is the reaction responsible for wine going sour when it is left exposed to the air.

Elimination reactions

Alcohols can also lose water (be dehydrated) leaving behind an alkene. This type of reaction is called an **elimination** as something – water in this case – is 'kicked out' of the molecule.

$$CH_3-CH_2-O-H \longrightarrow H_2C{=}CH_2 \;+\; H_2O$$

6 Classify the following reactions as addition, elimination, substitution, oxidation or dehydration.

```
     H  H  H  H                          H  H  H
     |  |  |  |                          |  |  |     H
a  H—C——C——C——C—O—H  ————————→        H—C——C——C=C/         + H2O
     |  |  |  |                          |  |      \
     H  H  H  H                          H  H       H

     H  H  H  H                          H  H  H  H
     |  |  |  |                          |  |  |  |
b  H—C——C——C——C  H  + H2  ————→      H—C——C——C——C—O—H
     |        |                          |  |  |  |
     H        H                          H  H  H  H

     H  H                               H
     |  |              [– 2H]           |   O
c  H—C——C—O—H        ————————→       H—C——C//
     |  |                               |    \
     H  H                               H     H

     H  Cl
     |  |                        H     H
d  H—C——C—H    ————————→          \   /
     |  |                          C=C            + HCl
     H  H                         /   \
                                 H     H

     H  H                                  H  H
     |  |                                  |  |
e  H—C——C—Br  + NaOH  ————————→        H—C——C—O—H   + NaBr
     |  |                                  |  |
     H  H                                  H  H
```

7 Write equations for the following reactions

a the oxidation of propan-1-ol
b elimination of water from propan-1-ol
c addition of bromine to propene
d addition of water to propene
e the reaction of propan-1-ol with hydrogen chloride
f the combustion of butan-1-ol. Hint: The butan-1-ol already has an oxygen atom of its own.

Name the organic products in each case.

Notice that in most of the reactions above, the length of the carbon chain is left unaffected. This is the pattern in the great majority of organic reactions.

Chapter 8

Energy changes in chemical reactions

STARTING POINTS

- Do you understand the following terms: **exothermic**, **endothermic**, **activation energy**? Write a sentence or two (no more) to explain what you understand by each term. Then check the glossary at the back of the book.

When chemical reactions take place they exchange energy (usually in the form of heat) with the surroundings. The name given to this heat energy exchange is the **enthalpy** change.

Enthalpy has the symbol H and units kJ/mol. We are usually more concerned with enthalpy *changes* which are given the symbol ΔH (pronounced 'delta-H'). Δ (the Greek letter delta) is used to represent a change in anything.

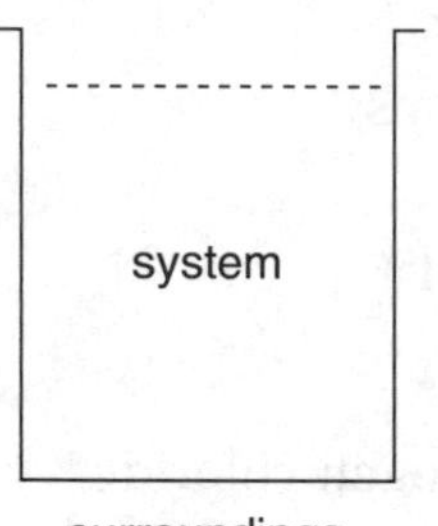

8.1 System and surroundings.

System is the name given to the reaction itself.

Surroundings includes everything outside – apparatus, atmosphere, etc.

Definition

The enthalpy change of a reaction is the energy exchanged with the surroundings at constant pressure.

The standard conditions are 1 atmosphere pressure and 298 K (25 °C).

If the energy exchange is from system to surroundings (heat is given out), the reaction is called **exothermic**, and we say that ΔH is negative.

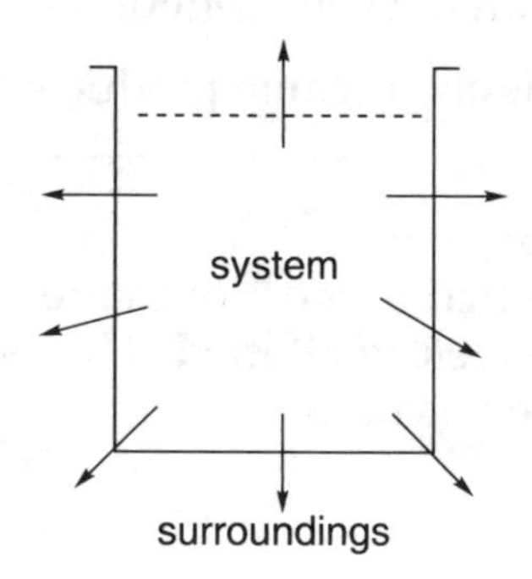

8.2 If energy in transferred from *system* to *surroundings* the reaction is **exothermic**.

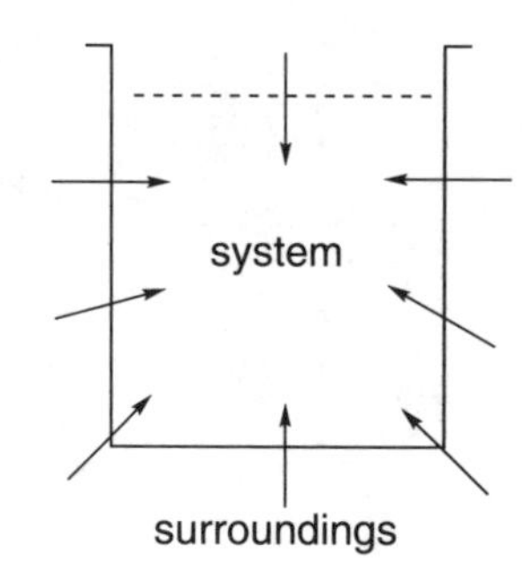

8.3 If energy in transferred from *surroundings* to *system* the reaction is **endothermic**.

If energy flows the other way, from surroundings into the system (heat is taken in), it is an **endothermic** reaction, and we say that ΔH is positive.

Energy level diagrams

A common way to show enthalpy changes is in an energy level (or, strictly, an enthalpy level) diagram. The two general diagrams are shown below.

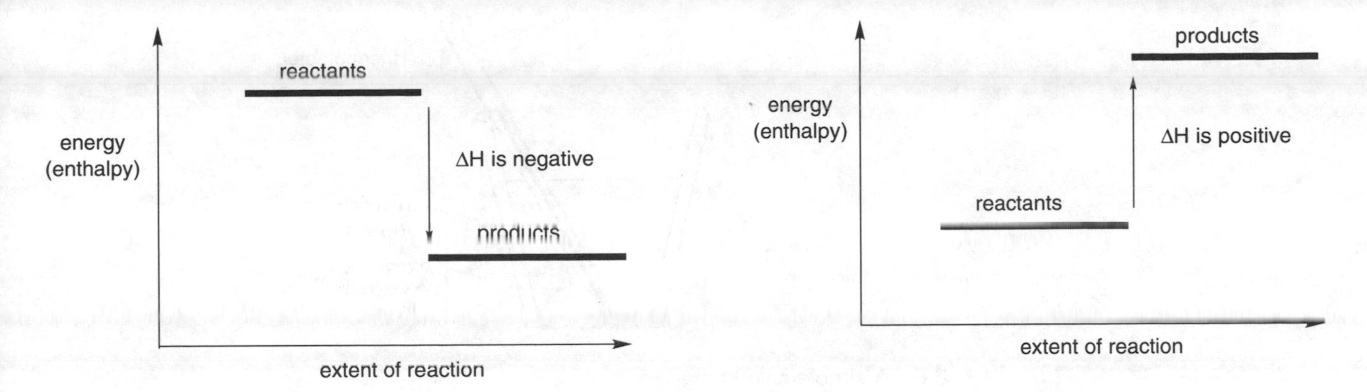

8.4 The energy level diagram for an exothermic reaction.

8.5 The energy level diagram for an endothermic reaction.

We normally replace the words reactants and products by the names of the substances concerned or, better, by a balanced symbol equation for the reaction. So for the reaction of hydrochloric acid with magnesium we could write

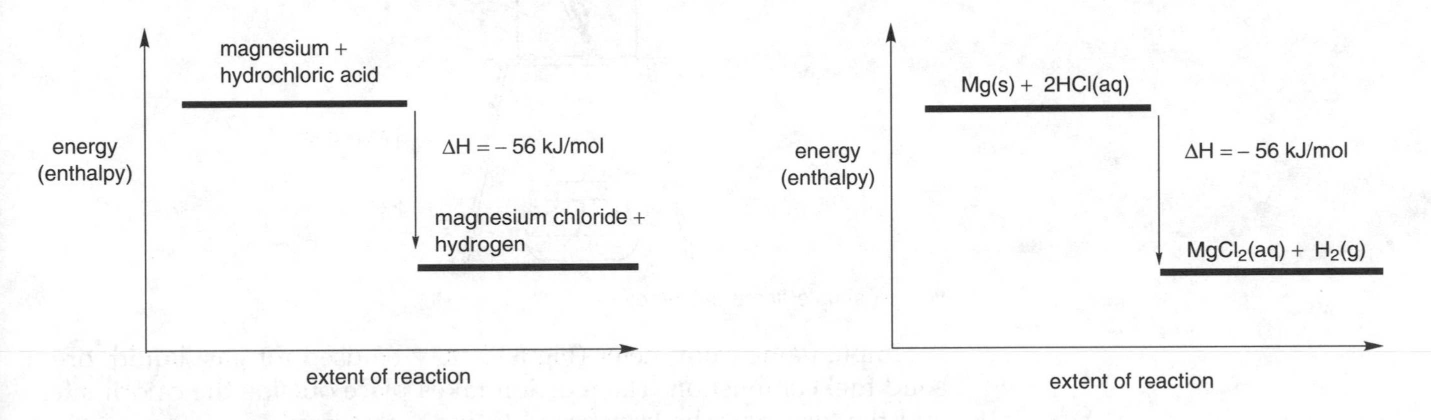

8.6 Energy level diagram for the reaction of magnesium with hydrochloric acid (words).

8.7 Energy level diagram for the reaction of magnesium with hydrochloric acid (symbols).

1 Draw a fully labelled energy level diagram for

a the burning of methane gas, CH_4, in oxygen ($\Delta H = -890$ kJ/mol)

b the photosynthesis reaction ($\Delta H = +2800$ kJ/mol)

$$6\,CO_2(g) + 6\,H_2O(l) \rightarrow C_6H_{12}O_6(s) + 6\,O_2(g)$$

Hint. The photosynthesis reaction stores up energy and is therefore endothermic.

Experimental methods

The two most common ways in which you are likely to have measured enthalpy change are by transferring the heat into a container of water – a simple calorimeter (Fig 8.8) or by using a simple flame calorimeter.

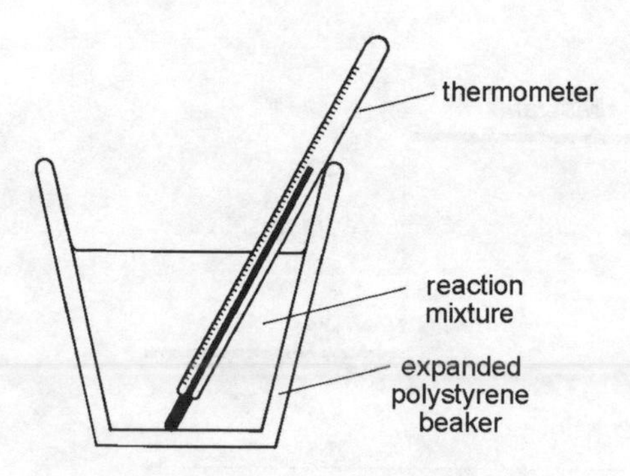

8.8 A simple calorimeter.

This can be used for displacement reactions, dissolving, or neutralisation. Here the reaction takes place in the calorimeter and the water of the solutions themselves is used to trap the heat.

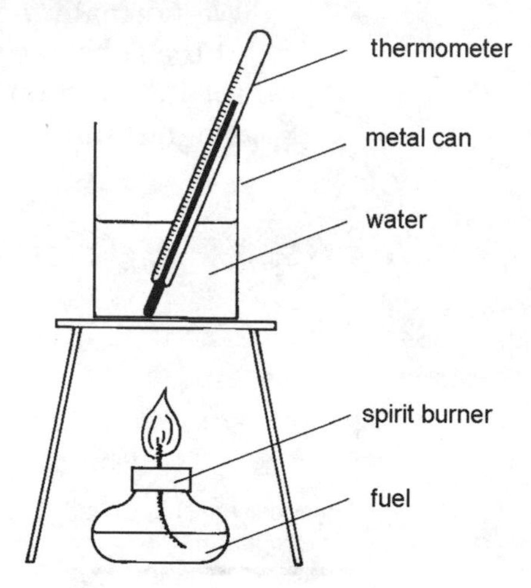

8.9 A simple flame calorimeter.

A simple flame calorimeter (Fig 8.9) may be used for gas, liquid, or solid fuel combustion. The reaction takes place outside the calorimeter and the heat must be transferred to the water inside.

Enthalpy calculations

The actual amount of the enthalpy change can be calculated from the temperature change of the water.

We need to know three things:

- the specific heat capacity of water, which is 4.2 J/g/°C (we usually count dilute solutions as though they were water)
- the mass of water being heated or cooled (again, we usually count dilute solutions as though they were water)
- the temperature change.

Written mathematically

Enthalpy change = 4.2 × mass of water × temperature change
(J) (J/g/°C) (g) (°C)

So if an exothermic reaction heats up 100 g of water by 5 °C

Enthalpy change = 4.2 × 100 × 5 = −2100 J

Or if an endothermic reaction cools down 50 g of water by 3 °C

Enthalpy change = 4.2 × 50 × 3 = +630 J

Note on units
We should measure the mass of water in grams but since one gram of water has a volume of 1 cm^3, we often measure the volume in cm^3 instead for convenience.

The temperature change is usually measured in °C. We can also use kelvin (K). Since K are the same size as °C, it will make no difference to the calculation. For example a temperature rise from 20 °C to 30 °C is 10 °C. In kelvin, this would be from 293 K to 303 K, still a rise of 10.

Note the signs – negative for an exothermic reaction and positive for an endothermic one.

ΔH is usually given in kJ/mol so the calculation is only complete when we divide our value for enthalpy change by 1000, converting J to kJ and then by the number of moles reacting, converting kJ to kJ/mol.

Signs

The signs of exothermic and endothermic reactions can be a little confusing. The easiest way to get them right is to remember that we think from the point of view of the chemicals, not ourselves. If you hold in your hand a test tube in which an exothermic reaction is taking place, your hand will get hot. This means that your hand is gaining energy and so the chemicals must be losing it. So, *from the point of view of the chemicals*, ΔH is negative. In an endothermic reaction, your hand will feel cold; heat is being transferred *from* your hand *into* the chemicals inside the test tube. The chemicals do not become hot because they store the energy in the form of chemical energy, not heat.

Instant cool packs

Sports trainers often keep instant cool packs in their first aid kits to treat sprains and bruises. These use endothermic reactions. One type consists of a thick plastic bag containing water and a thinner plastic bag of solid ammonium nitrate. To use the pack, the inner bag is broken with a sharp blow and the ammonium nitrate dissolves in the water – an endothermic reaction. A temperature drop almost sufficient to freeze the water can be produced.

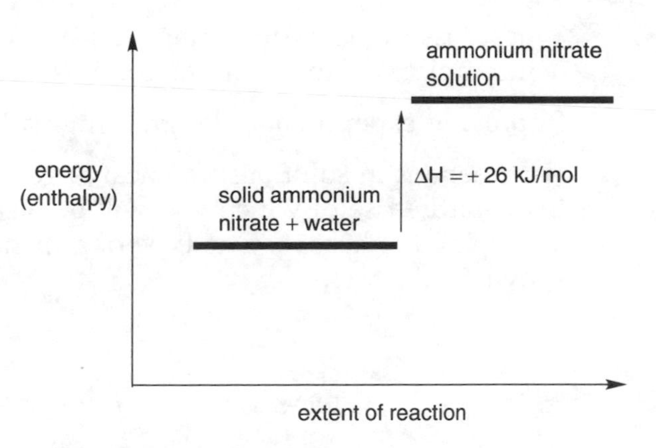

8.10 Energy level diagram for the reaction of solid ammonium nitrate with water.

An excess of zinc was added to 50 cm^3 0.2 mol/l copper sulphate solution. The temperature rose by 8 °C. Work out ΔH in kJ/mol.

We can take 50 cm^3 of dilute solution as having a mass of 50 g

$$\text{Enthalpy change} = 4.2 \times 50 \times 8 = 1680 \text{ J}$$

To work out the number of moles in this solution, use the formula

$$\text{Number of moles of solute} = M \times \frac{V}{1000}$$

$$= 0.2 \times \frac{50}{1000} = \frac{10}{1000}$$

$$= 0.01 \text{ mol}$$

Problem with calculating moles? See Chapter 2 on Chemical calculations.

0.01 mol produces 1680 J of heat so

$$1 \text{ mol will produce } \frac{1680}{0.01} = 168\ 000 \text{ J}$$

$$\text{This is } \frac{168\,000}{1000} = 168 \text{ kJ/mol}$$

$\Delta H = -168$ kJ/mol. The negative sign is because heat is given out in the reaction.

The signs and units are important. Leaving them out is bad practice and will cost you marks in examinations.

Now work out ΔH for the following examples.

2 An excess of magnesium was added to 100 cm^3 of 0.5 mol/l copper(II) sulphate solution. The temperature rise was from 16 °C to 56 °C.

3 0.02 moles of ethanol burned to produce a temperature rise of 13 °C in 200 g water in a beaker.

Eliminating errors

In question 3 the value you calculated for ΔH should have been −546 kJ/mol.

The accepted value in a data book is −1367 kJ/mol.

The difference is due to experimental errors. There are two main factors:

- heat loss to the surroundings (for all experiments)
- incomplete combustion (for combustion only).

Improving experimental design can greatly reduce the errors.

For reactions in solutions, the heat is generated in the water the temperature rise of which you will be measuring. The problem here is keeping the heat in. One type of apparatus you might use is shown in Fig 8.11.

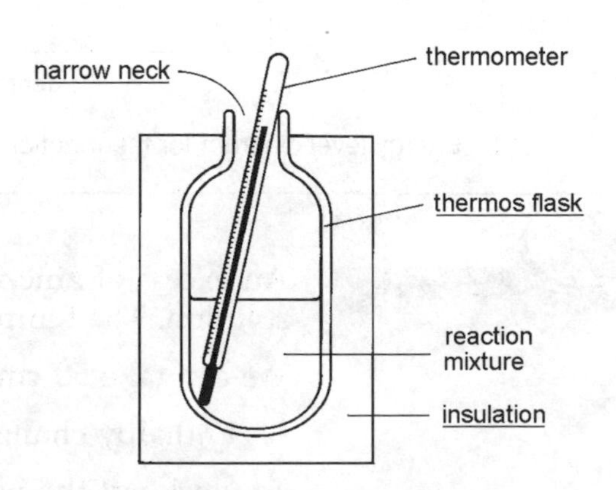

8.11 An improved simple calorimeter.

4 Explain how each of the underlined labelled features of the apparatus shown in Fig 8.11 keeps the heat in.

For combustion reactions, there are three problems:

- getting the heat from the flame into the water
- keeping the heat in the water once it is there
- ensuring that all the fuel burns completely.

If there is a shortage of oxygen, fuels will not burn completely and less energy will be given out. For example carbon-containing fuels might burn to give carbon monoxide or carbon rather than carbon dioxide. Unburnt carbon can often be seen as 'soot' on the apparatus. A yellow bunsen flame is an example of incomplete combustion.

The apparatus shown in Fig 8.12 can be used to reduce experimental errors in combustion experiments. It is called a flame calorimeter.

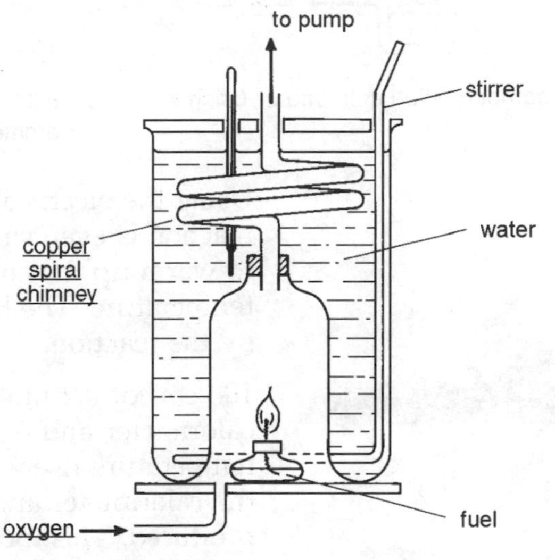

8.12 An improved flame calorimeter.

5 List the ways that the underlined labelled features of the apparatus in Fig 8.12 improve experimental accuracy.

Electrical calibration

Electrical calibration gives us an alternative way of measuring enthalpy changes. For an exothermic reaction, we use an electrical heater to heat up the reaction mixture and apparatus by the same amount as it is heated up by the reaction. The heat produced by the reaction must have been the same as the heat produced by the electric heater (including any heat lost from the apparatus, which will be the same in both parts of the experiment). We measure the energy given out by the heater electrically in one of two ways. We can either connect it to a joulemeter (rather like the electricity meter at home) which measures the energy directly (see Fig 8.13) or we can measure the current flowing through the heater, the voltage across the heater and the time for which it is on (see Fig 8.14). We can then use the expression

Electrical energy supplied $= V \times I \times t$

where

V is the potential difference (or voltage) across the heater in volts.

I is the current flowing through the heater in amps

t is the time for which the heater is on in seconds.

We can also use the electrical heater method with the apparatus shown in Fig 8.11.

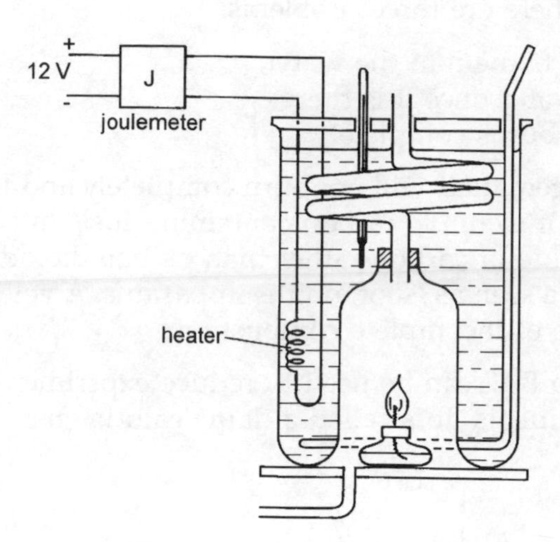

8.13 An electrical compensation calorimeter using a joulemeter.

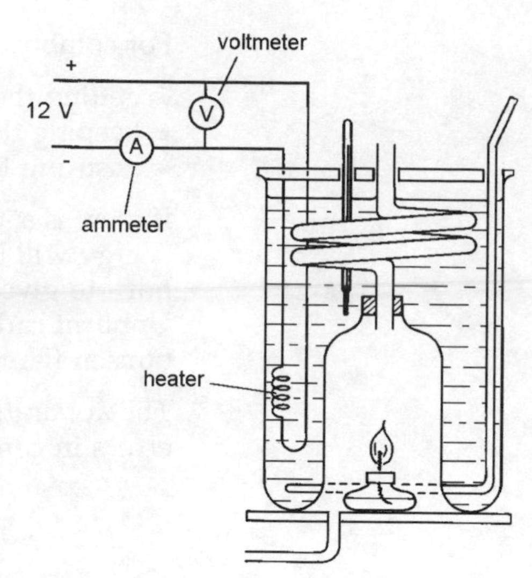

8.14 An electrical compensation calorimeter using a voltmeter and ammeter.

Using the electrical compensation method for an endothermic reaction is even simpler. We carry out the reaction and use the heater to warm up the cooled reaction mixture and apparatus to its original temperature. The heat energy required to do this is the heat absorbed by the reaction.

100 cm^3 of 0.2 mol/l copper sulphate solution was placed in a calorimeter and 6.5 g (an excess) of powdered zinc added. The temperature rise was measured and an electrical heater was placed in the calorimeter and switched on until the same temperature rise was produced. This took 30 minutes with a current of 0.2 A and a voltage of 12 V. What is ΔH for the reaction?

$$\text{Electrical energy supplied} = V \times I \times t$$
$$= 12 \times 0.2 \times (30 \times 60)$$
$$= 4320 \text{ J}$$

$$\text{Number of moles of copper sulphate} = M \times \frac{V}{1000}$$
$$= 0.2 \times \frac{100}{1000}$$
$$= 0.02 \text{ mol}$$

0.02 mol of copper sulphate produces 4320 J so

$$1 \text{ mol would produce } \frac{4320}{0.02} = 216\,000 \text{ J}$$

This is −216 kJ/mol, the negative sign shows that heat is given out.

6 0.046 g of ethanol was burned in a flame calorimeter and produced a temperature rise of 4 °C. An electrical heater connected to a joulemeter required 1300 J to produce the same temperature rise. What is ΔH for the reaction?

7 100 cm^3 of 0.2 mol/l copper sulphate solution was placed in a calorimeter and 2.4 g (an excess) of powdered magnesium added. The temperature rise was measured and an electrical heater was placed in the calorimeter and switched on until the same temperature rise was produced. This took 70 minutes with a current of 0.2 A and a voltage of 12 V. What is ΔH for the reaction?

The most accurate work is done using a bomb calorimeter.

The bomb calorimeter

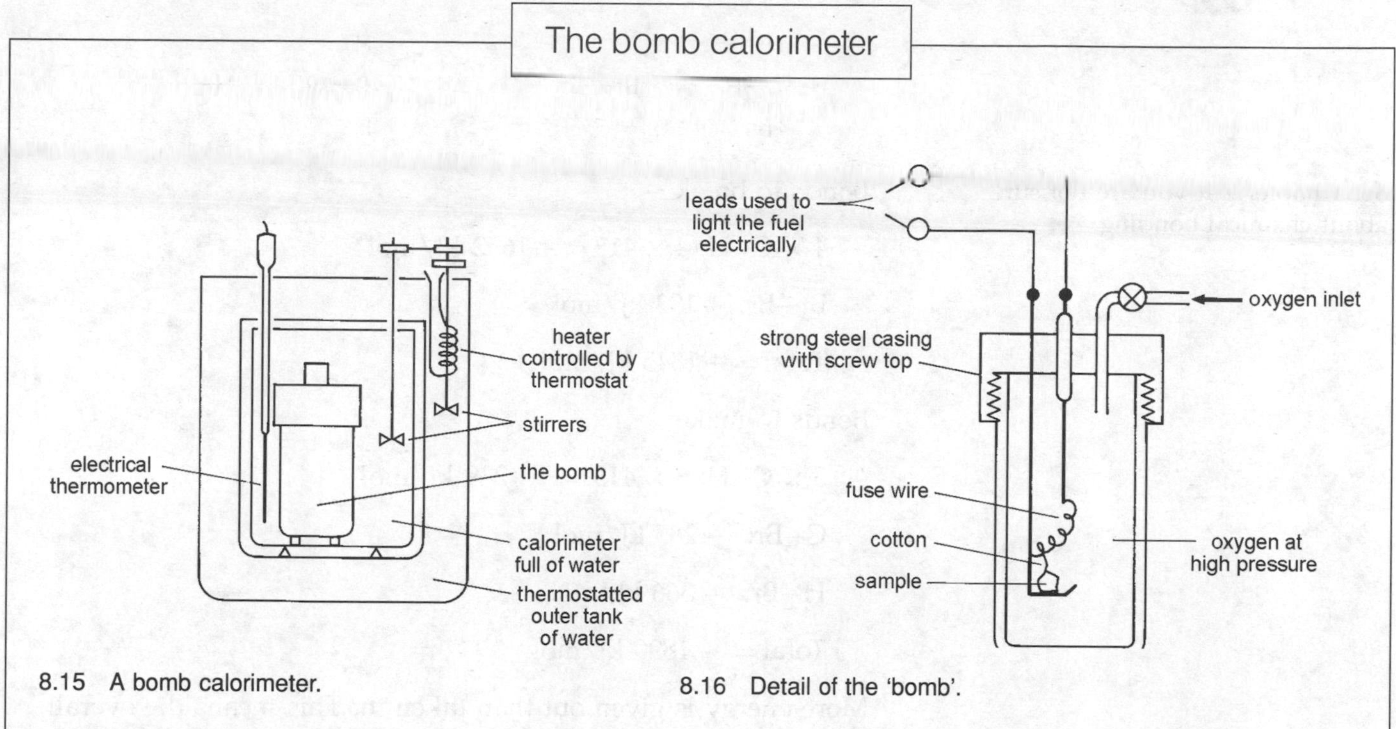

8.15 A bomb calorimeter.

8.16 Detail of the 'bomb'.

This is how accurate data book values of ΔH are determined. The 'bomb' is a stainless steel vessel which holds the sample and can be filled with oxygen at high pressure to ensure complete combustion. Heat loss from the calorimeter is eliminated by keeping the thermostatically controlled outer tank of water at the same temperature as the water in the calorimeter.

We determine what is called the heat capacity of the bomb by burning benzoic acid whose enthalpy change of combustion is accurately known. The heat capacity is the amount of heat needed to raise the temperature of the whole calorimeter by 1 °C.

Readings are taken to a high degree of precision; mass to ± 0.0001 g and temperature to ± 0.001 °C.

Table 8.1 Selected bond enthalpies

Bond	Bond enthalpy/ kJ/mol
H–H	436
O–O	144
O=O	498
O–H	464
C–H	413
C–C	347
C=C	612
C–O	358
C=O	750
C–F	452
C–Cl	346
C–Br	285
Cl–Cl	243
Br–Br	193
H–Cl	432
H–Br	366

Bond enthalpy

Bond enthalpies (often called bond energies) can be used to calculate approximate ΔH values or reactions.

When bonds are *broken*, energy has to be *put in* to break the bond. The amount depends on the strength of the bond – see Table 8.1.

So the amount of energy needed to **break** the bonds in CO_2, O=C=O, (leaving a separate carbon atom and two separate oxygen atoms) is

$$2 \times 750 = +1500 \text{ kJ/mol}$$

When bonds are *made*, energy is given out. So making water, H–O–H, from one oxygen and two hydrogen atoms releases

$$2 \times 464 = -928 \text{ kJ/mol}$$

The energy difference between breaking existing bonds and making new bonds determines whether a reaction is endothermic or exothermic overall. If more energy is given out than is taken in, the reaction is exothermic and if more energy is taken in than is given out, it is endothermic.

A methane + bromine → bromomethane + hydrogen bromide

$CH_4 + Br_2 \rightarrow CH_3Br + HBr$

H–C(H)(H)–H + Br–Br ⟶ H–C(H)(H)–Br + H–Br

See Chapter 2 if you are not sure about chemical bonding.

Bonds to break

4 × C–H: 4 × 413 = +1652 kJ/mol

Br–Br: +193 kJ/mol

Total = +1845 kJ/mol

Bonds to make

3 × C–H: 3 × 413 = – 1239 kJ/mol

C–Br: – 285 kJ/mol

H–Br: – 366 kJ/mol

Total = – 1890 kJ/mol

More energy is given out than taken in. This means the overall balance is – 45 kJ/mol released, i.e the reaction is exothermic.

B methane + oxygen → carbon dioxide + water

$CH_4 + 2O_2 \rightarrow CO_2 + 2H_2O$

H–C(H)(H)–H + O=O, O=O ⟶ O=C=O + H–O–H, H–O–H

It is a good idea to draw the **displayed** formulae (see Chapter 7) in bond energy calculations so that you do not miss any bonds. Remember that if there are, say, two moles of oxygen then you will have to break two moles of bonds.

Bonds to break

4 × C–H: 4 × 413 = +1652 kJ/mol

2 × O=O: 2 × 498 = +996 kJ/mol (the 2 is because there are two moles of oxygen)

Total = +2648 kJ/mol

Bonds to make

2 × C=O: 2 × 750 = –1500 kJ/mol

4 × O–H: 4 × 464 = –1856 kJ/mol (the 4 is because there are two moles of water each with two O–H bonds)

Total = –3356 kJ/mol

More energy is given out than taken in. This means the overall balance is – 708 kJ/mol released, i.e. the reaction is exothermic.

Now work out ΔH for the following reactions

8 $C_5H_{12} + 8O_2 \rightarrow 5CO_2 + 6H_2O$

9 $C_3H_8 + 5O_2 \rightarrow 3CO_2 + 4H_2O$

10 $C_2H_5OH + 3O_2 \rightarrow 2CO_2 + 3H_2O$

Chapter 7 on Organic chemistry will help you with bonds in carbon compounds.

A short cut

In many reactions, most of the bonds stay intact and are neither broken nor made. You can often simplify bond energy calculations by identifying these bonds and leaving them out of your calculations.

$$CH_4 + Cl_2 \rightarrow CH_3Cl + HCl$$

$$\text{H–C(H)(H)–H} + \text{Cl–Cl} \longrightarrow \text{H–C(H)(H)–Cl} + \text{H–Cl}$$

The only bonds that break are one of the C–H bonds and the Cl–Cl bond. The only bonds that are made are one C–Cl bond and one H–Cl bond. All the others stay intact.

Bonds to break

C–H: + 413 kJ/mol

Cl–Cl: +243 kJ/mol

Total = +656 kJ/mol

Bonds to make

C–Cl: – 346 kJ/mol

H–Cl: – 432 kJ/mol

Total = – 778 kJ/mol

So the difference is – 122 kJ/mol

11 Try the calculation in the worked example above using the method of breaking and making all the bonds. You should get exactly the same answer.

12 Work out ΔH for the following reactions by considering only the bonds which are actually made and broken.

a $C_2H_5OH + HCl \rightarrow C_2H_5Cl + H_2O$
b $CH_2CH_2 + Cl_2 \rightarrow CH_2ClCH_2Cl$

13 Plot a graph of ΔH of combustion for methane (CH_4), propane (C_3H_8) and pentane (C_5H_{12}) against the number of carbon atoms in the molecule. How much does ΔH go up each time? What would you predict for ΔH of combustion of heptane (C_7H_{16})?

You should be able to see that there is a pattern in ΔH for the combustion of the alkanes. The regular increase in ΔH is a reflection of the pattern in bond breaking. As an extra carbon atom is added to the chain in an alkane we are adding the group H–C–H, i.e. 1 C–C bond + 2 C–H bonds.

```
    H            H  H
    |            |  |
 H—C—H       H—C—C—H
    |            |  |
    H            H  H
```

The extra atoms require $1\frac{1}{2}$ molecules of O_2 to react with and convert them into one molecule of CO_2 and one of H_2O.

So we must break

$1\frac{1}{2}$ O=O: $1\frac{1}{2} \times 498 = +747$ kJ/mol

1 × C–C: +347 kJ/mol

2 × C–H: 2 × 413 = +826 kJ/mol

Total = +1920 kJ/mol

and make

2 × C=O: 2 × 750 = – 1500 kJ/mol

2 × O–H: 2 × 464 = – 928 kJ/mol

Total = – 2428 kJ/mol

This is an extra 508 kJ/mol of energy given out for each extra CH_2 group. Compare this with the figure you worked out from the graph in question 13.

The strengths of bonds

Bond energies make it possible to predict *which* bond is most likely to break in a compound as the bond with the smallest bond energy is the easiest to break. We need to take care as other factors can affect bond breaking, but this method is a useful guide.

Which is the most likely bond to break in bromoethane, H–C(H)(H)–C(H)(H)–Br ?

Look up the bond energies in Table 8.1 and write them next to the bonds on the displayed formula of bromoethane.

```
      H       H
   413|       |413
 H————C——347——C————Br
   413|       |  285
   413|       |413
      H       H
```

You should be able to see that the C–Br bond is the weakest. This is in fact the bond that breaks in most of the reactions of bromoethane.

14 Predict the bond most likely to break in the following molecules.

a CH_3Cl
b CH_3F
c $C_2H_5OOC_2H_5$

Hardening fibreglass

Fibreglass, used for repairing car bodies and making canoes, is a mat of woven glass threads held together by a polymer resin. This resin is applied to the glass as a liquid containing small molecules (monomers) and a hardener is added which starts a reaction which joins these monomers to make the solid resin. The monomer contains the molecule benzoyl peroxide whose bond energies are shown.

Look at Table 8.1 on p. 63. You can see that the O–O bond is very weak compared with the others. Even at room temperature, this bond can break, splitting the molecule in half and leaving two highly reactive fragments called radicals. These are what start the polymerisation process which hardens the resin.

Activation energy

Bond enthalpies help to explain why many reactions which take place very readily need energy to start them off. This energy is called the **activation energy**. Petrol, for example, can be stored quite safely in contact with oxygen in the air but if a spark or flame is applied it will burn rapidly. The energy of the spark or flame is needed to start breaking bonds in the reactants. The stronger the bonds in the reactants, the greater the activation energy required. Since the reaction is exothermic, more energy is produced when the bonds form in the products and this energy can begin to break more bonds in the reactants. So, once started, the reaction continues. The reaction has an energy 'hill' to climb before it can start. This is shown on the energy level diagram in Fig 8.17 where the symbol E_A is used for activation energy.

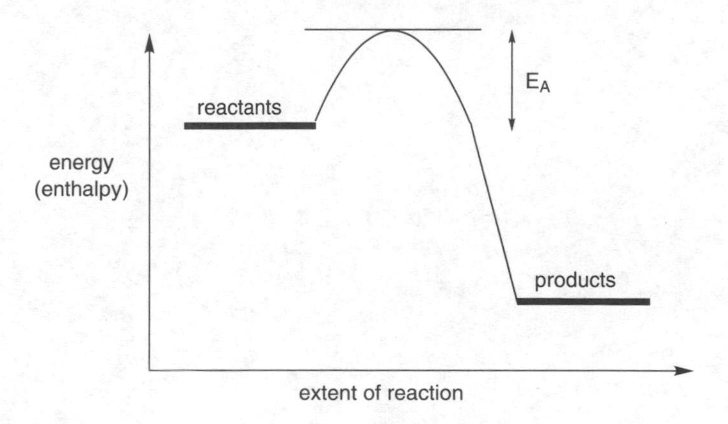

8.17 Activation energy.

15 Draw the energy diagram, including E_A, for an endothermic reaction.

Catalysts

Catalysts work by providing alternative, lower energy pathways for a reaction. In other words, they reduce the activation energy for the reaction in some way.

For example the reaction

$$2N_2O \rightarrow 2N_2 + O_2$$

has an activation energy of 240 kJ/mol without a catalyst but only 120 kJ/mol with a gold catalyst (see Fig 8.18).

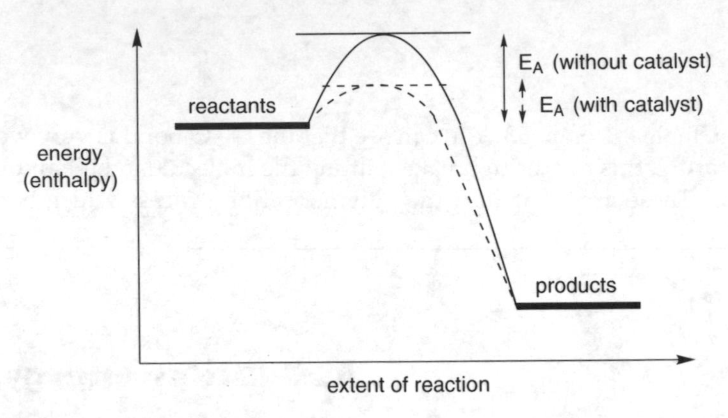

8.18 Catalysts and activation energy.

16 Suggest what energy source provides the activation energy for the following reactions:

a lighting a match
b petrol burning in an internal combustion engine
c photosynthesis

Chapter 9 Electrolysis

STARTING POINTS

- You should understand: simple circuit electricity, atomic structure, ionic bonding, metallic bonding, molecules.

When an electric current passes through a solid, the solid is unchanged by the current. Wires that carry a current may get warm but they stay the same substance. A solid which conducts electricity is called a good conductor.

1 What sorts of solids are in general good conductors of electricity? Explain why.

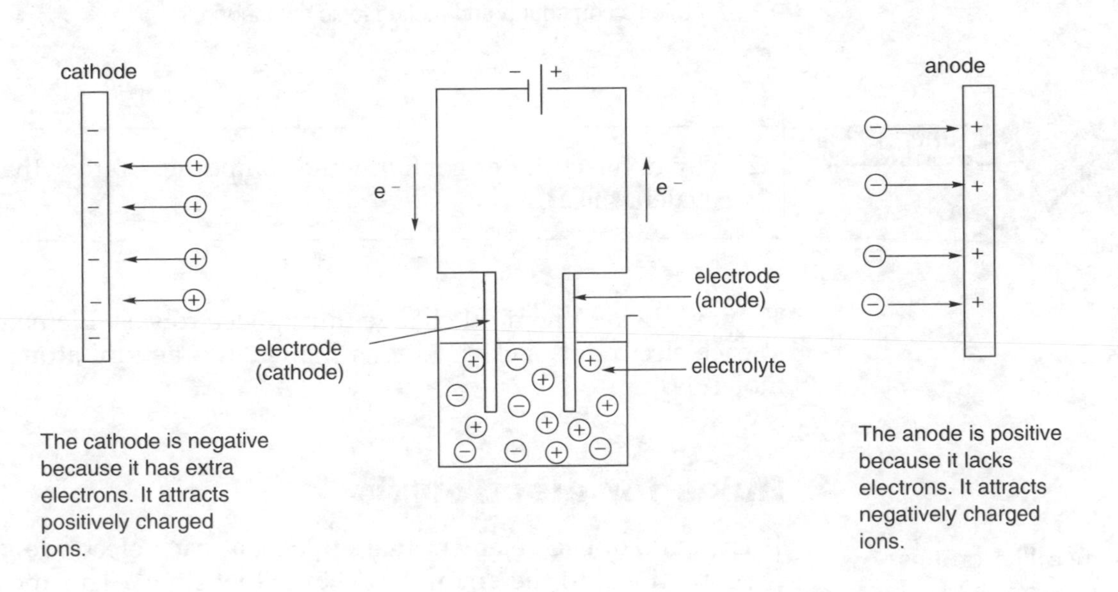

9.1 An electrolysis cell.

A liquid or solution which will conduct electricity and is chemically changed in the process is called an **electrolyte**. When a current passes through an electrolyte, chemical changes occur at the **electrodes**. This process is called **electrolysis**.

Some chemical history

Electrolysis is a process that we take for granted now, but when it was first discovered it was a major leap forward for chemistry. This was because it was the only way to unlock compounds which contained reactive metals like sodium and calcium. Before electrolysis, scientists assumed that calcium oxide, then called 'calcite', for example, was an element, because it couldn't be broken down into anything simpler. Sir Humphrey Davy used electrolysis to discover the elements sodium, potassium, calcium, magnesium, barium and strontium.

What is a current?

9.3 Electrolysis is used to plate one metal with another.

A current is a flow of charge. In a metal conductor it is the negative electrons that carry charge. In an electrolyte, the charge is carried by the movement of positive and negative ions. Remember that ions are only free to move when an ionic solid is melted or dissolved in water. Ionic *solids* cannot conduct because the ions are too strongly attracted to each other to move apart.

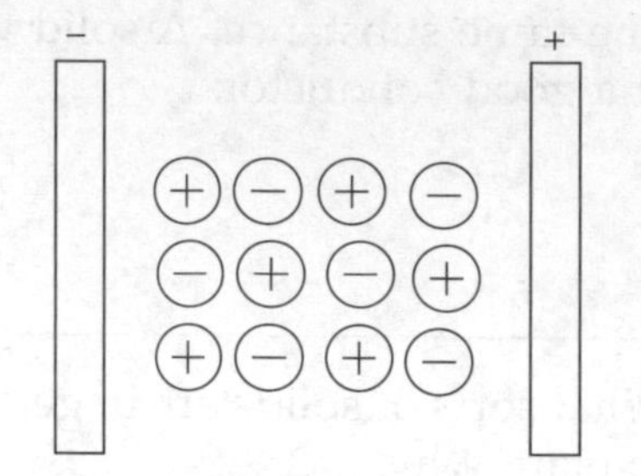

Solid ionic compound will not conduct electricity.

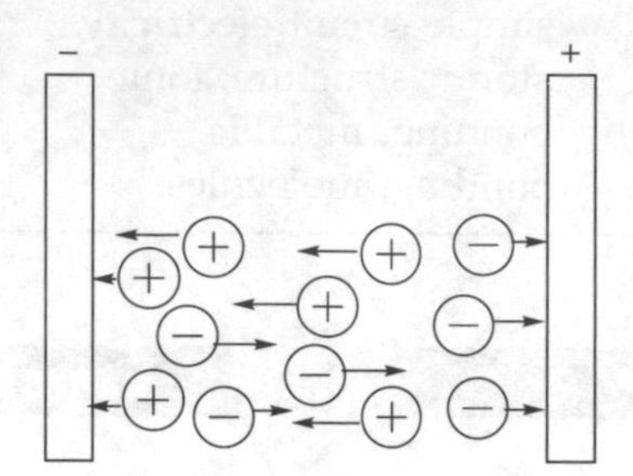

In the liquid state ions can move.

9.2 Solid ionic compound and melted ionic compound.

2 Why do electrolytes contain ionic compounds rather than covalent ones?

There is always chemical change during electrolysis. Elements appear at each electrode, as charged ions change into neutral atoms or molecules.

Rules for electrolysis

Positive ions are called **cations** because they are attracted to the cathode. This phrase might help – 'copper **c**ations at the **c**athode' – if you can remember that copper forms Cu^{2+} (positive) ions. Negative ions are called **anions** because they are attracted to the positive anode. Remember the word **an**ode contains **and** which means plus.

1 The mass of each element that appears at each electrode is proportional to the amount of charge that flows. This means for example that as we double the amount of charge, so the mass discharged at the electrode doubles (see Chapter 11 on Mathematics if you are not sure about proportionality). You would probably predict this, but what is interesting is that for the same amount of electricity, the actual mass varies from element to element. You will see why later in the chapter.
2 Metal elements or hydrogen appear at the cathode. This is because metals and hydrogen form positive ions. These ions are attracted to the negative cathode and gain from it the number of electrons that will make them neutral.
3 Non-metals appear at the anode. This is because non-metals form negative ions and these are attracted to the positive anode. Negative ions lose to the anode the number of electrons that will make them neutral.

We can write equations for the processes which occur at the electrodes.

For example, we can electrolyse liquid (melted) sodium chloride, NaCl (see Fig 9.4).

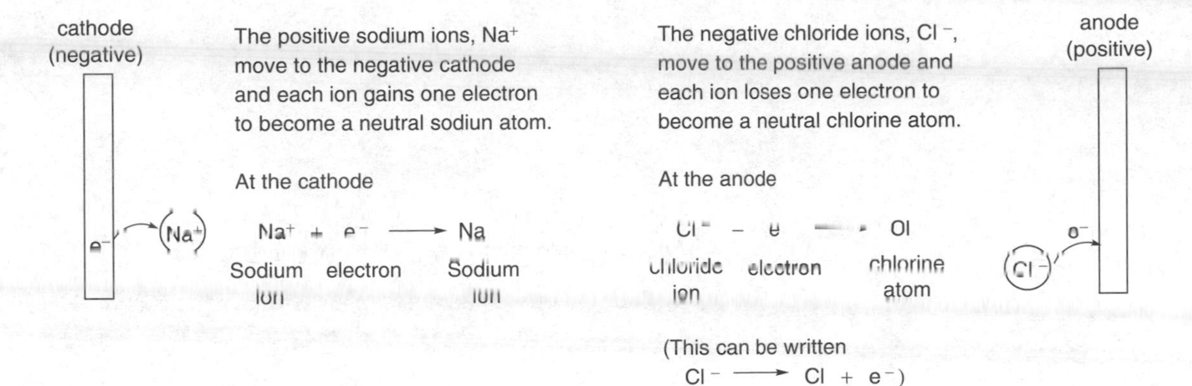

9.4 Electrolysis of molten sodium chloride.

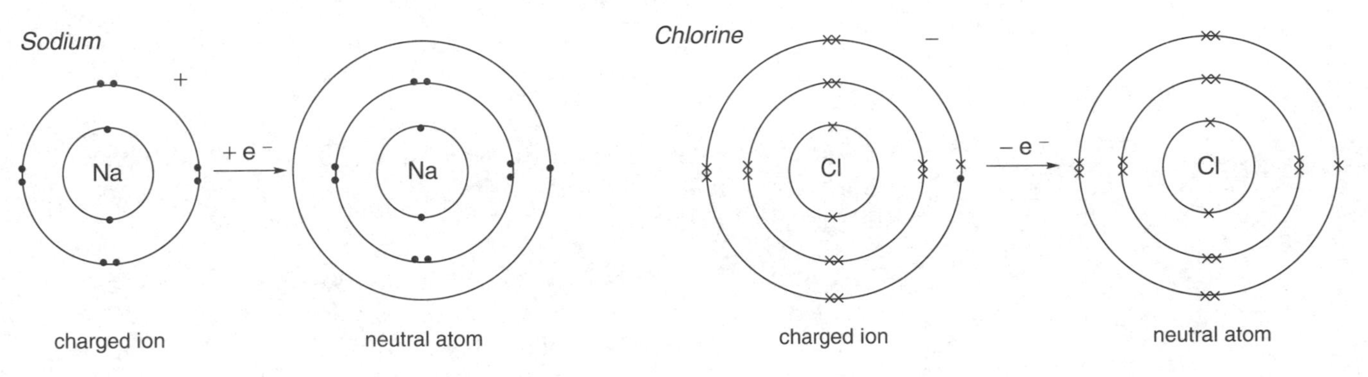

9.5 Neutralisation of sodium and chloride ions.

Reactions at the cathode and anode always happen at the same time and exactly the same number of electrons are gained at the cathode as are lost at the anode but we often take them separately and call them **half equations**.

Cathode reactions

All metal ions gain from the cathode the same number of electrons as their ionic charge. So a metal ion with a charge of 1+ gains one electron

$$M^+ + e^- \rightarrow M$$

a metal ion with a charge of 2+ gains two electrons

$$M^{2+} + 2e^- \rightarrow M$$

and a metal ion with a charge of 3+ gains three electrons

$$M^{3+} + 3e^- \rightarrow M$$

(Metals do not usually have ions with charges of more than 3+.) It is useful to remember that the charge on the ion is the same as the Group number in the Periodic Table for metals in Groups 1, 2 and 3.

Anode reactions

Similarly, negative ions will lose the same number of electrons as their charge. It is useful to know that the charge on the ion is the Group number minus 8 for Groups 7, 6 and 5. For example, oxygen is in Group 6 so the oxygen ion will have the charge

$$6 - 8 = -2, \text{ i.e. } O^{2-}$$

Group	1	2	3	4	(5 – 8) 5	(6 – 8) 6	(7 – 8) 7	(8 – 8) 8
Charge	1+	2+	3+	–	3–	2–	1–	0

Note: It takes too much energy to gain or lose 4 electrons

9.6 Charges on ions can be worked out from the Group number in the Periodic Table.

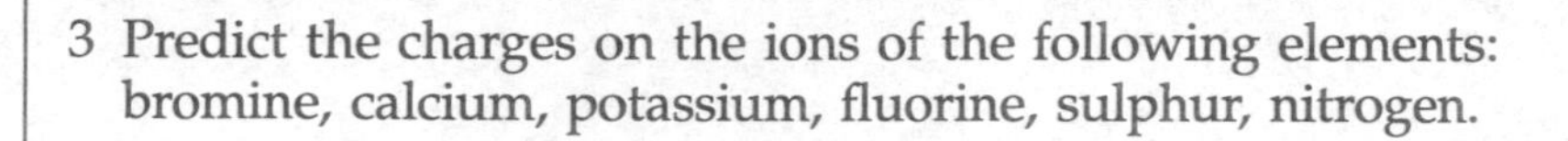

3 Predict the charges on the ions of the following elements: bromine, calcium, potassium, fluorine, sulphur, nitrogen.

A Molten magnesium chloride, $MgCl_2$, is electrolysed. What are the products at each electrode? Write half equations for the reactions at the electrodes.

Answer: It is always easy to predict the result of electrolysing a molten compound because there usually isn't any choice. The products here are the metal, magnesium, at the cathode and the non-metal, chlorine, at the anode.

Gases (except the inert gases) are not stable as atoms so, as soon as an atom is discharged, it will immediately combine to form a molecule.
For example

$$Cl^- \rightarrow Cl + e^-$$

which then leads to

$$2\,Cl \rightarrow Cl_2$$

Helpful hint. It is often easier to write the whole process as:

$$Cl^- \rightarrow \tfrac{1}{2}Cl_2 + e^-$$

This is perfectly correct because it means that for each mole of chloride ions we get $\frac{1}{2}$ mole of chlorine molecules.

Magnesium is in Group 2 so the ions have a charge of 2+

Chlorine is in Group 7 so the ions have a charge of –1 (7 – 8).

At the cathode

$$Mg^{2+} + 2\,e^- \rightarrow Mg$$

At the anode

$$Cl^- \rightarrow Cl + e^-$$

or

$$Cl^- \rightarrow \tfrac{1}{2}Cl_2 + 2\,e^-$$

B Aluminium oxide, Al_2O_3, is melted and electrolysed. What are the products at each electrode? Write the half equations for the reactions at the electrodes.

Answer: Aluminium will form at the cathode. Oxygen will form at the anode.

Aluminium is in Group 3 and its ion has a charge of 3+. Oxygen is in Group 6 and an oxide ion has a charge of 2 – (6 – 8).

At the cathode

$$Al^{3+} + 3\,e^- \rightarrow Al$$

At the anode

$$O^{2-} \rightarrow O + 2\,e^-$$

or

$$O^{2-} \rightarrow \tfrac{1}{2}O_2 + 2\,e^-$$

and to get the balance right, we need to involve the same number of electrons, so:

$$2\,Al^{3+} + 6\,e^- \rightarrow 2\,Al \qquad 3\,O^{2-} \rightarrow 1\tfrac{1}{2}O_2 + 6\,e^-$$

4 State the products and write the half equations for the result of electrolysing the following molten compounds:

a calcium oxide
b aluminium chloride
c lead(II) bromide (II means that the lead ion is 2+)

Electrolysing aqueous solutions

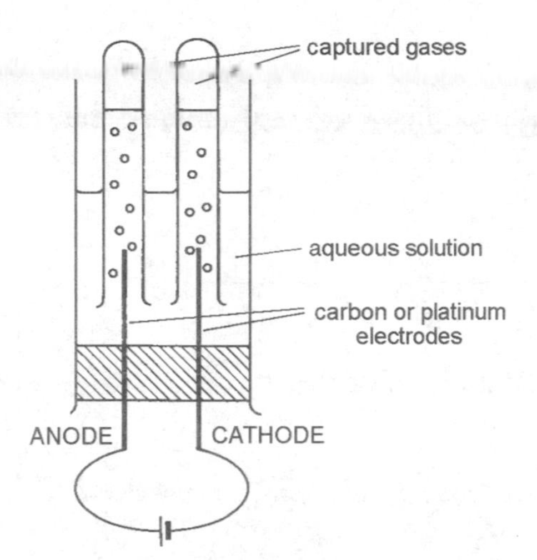

9.7 An electrolysis cell to capture gases.

Dissolving compounds in water to electrolyse them (making an aqueous solution) is easier than melting them. However the presence of water usually makes a difference to the products.

This is because although water has covalent bonding and is therefore a poor conductor of electricity there is always a small number of hydrogen ions ($H^+(aq)$) and hydroxide ions ($OH^-(aq)$) present ((aq) means dissolved in water).

$$\underset{\text{molecule}}{H_2O} \rightleftharpoons \underset{\text{ion}}{H^+(aq)} + \underset{\text{ion}}{OH^-(aq)}$$

At the cathode

In an aqueous solution of a metal salt both metal ions and hydrogen ions are present. The reactive metals, like sodium and calcium, stay as ions (after all reactive metals usually exist as ions – that's what it means to be reactive). So, the hydrogen ions then 'win' the electrons and only hydrogen gas appears.

5 Write the equation for the formation of hydrogen gas at the cathode. Remember that hydrogen gas exists as H_2 molecules and not atoms.

Unreactive metals, like silver and copper, will appear instead of hydrogen.

Other factors that can change the products of electrolysis are the concentration of the solution and the electrodes used; in industry the exact conditions of electrolysis are often a secret.

At the anode

Negative ions are often complex (that is, they contain more than one atom, e.g. sulphate ion, SO_4^{2-}), but the outcome is simpler.

Oxygen will usually appear at the anode. This comes from the reaction of the hydroxide ion at the anode.

$$4OH^-(aq) \rightarrow O_2(g) + 2H_2O(l) + 4e^-$$

But when halide ions – $Cl^-(aq)$, $Br^-(aq)$, $I^-(aq)$, but *not* $F^-(aq)$ – are present in sufficient concentration, they will appear because there are so very many of them.

A Concentrated sodium chloride solution (brine) is electrolysed in industry and produces useful products that are the starting materials for other chemicals. Name *three* useful products.

Answer: The ions present in brine are $Na^+(aq)$, $H^+(aq)$, $OH^-(aq)$, and $Cl^-(aq)$.

At the cathode we get hydrogen rather than sodium, because sodium is a reactive metal.

At the anode we get chlorine, because the solution is concentrated.

This leaves behind in the solution $Na^+(aq)$ and $OH^-(aq)$ which is sodium hydroxide.

So the three useful products obtained from the electrolysis of brine are hydrogen, chlorine and sodium hydroxide.

B What would be the result of electrolysing a solution of magnesium sulphate?

Answer: The ions present are $Mg^{2+}(aq)$, $H^{+}(aq)$, $SO_4^{2-}(aq)$, $OH^{-}(aq)$.

Magnesium is a reactive metal and stays in solution so we get hydrogen instead at the cathode.

Sulphate ions stay in solution and we get oxygen at the anode.

This leaves behind the $Mg^{2+}(aq)$ and the $SO_4^{2-}(aq)$ ions, so the solution gradually becomes more concentrated. In fact we have simply electrolysed the water.

6 Predict the result of electrolysing:

a a very dilute solution of potassium chloride
b a solution of calcium nitrate
c dilute sulphuric acid

The electrodes

The electrodes we use can also make a difference. For example, if we electrolyse copper sulphate solution using copper electrodes we get copper at the cathode as we might predict, but this time the copper anode dissolves into solution and oxygen is not produced. This is used as the method for purifying copper in industry.

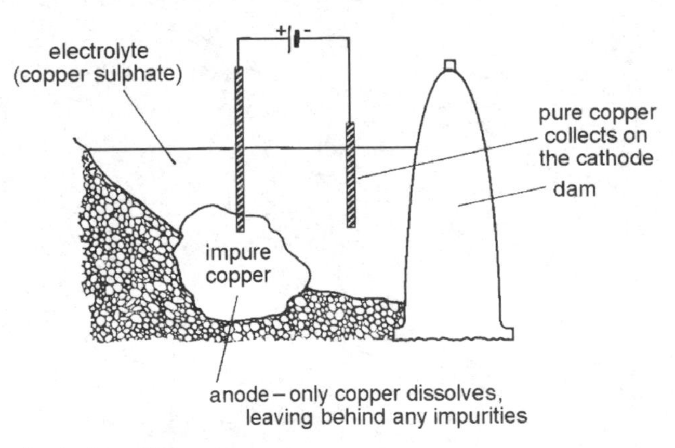

9.8 One method of purifying copper.

Here are the half equations for this process

At the cathode

$$Cu^{2+}(aq) + 2e^- \rightarrow Cu(s)$$

At the anode

$$Cu(s) \rightarrow Cu^{2+}(aq) + 2e^-$$

(the anode itself)

Electrolysis calculations

Electrolysis produces elements at the electrodes, but the same amount of charge will not produce the same mass of different elements. This is because the mass of the element that is discharged depends on the mass of a mole of the element.

To work out the mass, we use the technique of moles with half equations in the same way that we used them for the more usual equations. For example:

Cu^{2+}	+	$2e^-$	$\rightarrow$	Cu
1 mole		2 moles		1 mole

This tells us that if 1 mole of copper ions gain 2 moles of electrons we get 1 mole of copper deposited – but what is a mole of electrons?

The faraday

Electric charge is measured in coulombs. Electrons carry charge. Each electron carries 1.6×10^{-19} coulombs. So 1 mole of electrons (6×10^{23}) carries a charge of $1.6 \times 10^{-19} \times 6 \times 10^{23} = 96\,000$ coulombs. This amount of charge is given the name 1 faraday (F). This might seem difficult, but it is easy to use.

1 mole of electrons = 1 faraday = 96 000 coulombs

We can now add this to our equation:

Cu^{2+}	+	$2e^-$	$\rightarrow$	Cu
1 mole		2 moles		1 mole
64 g		2 faradays		64 g (we ignore the tiny mass of electrons)
		$2 \times 96\,000$ coulombs		
		192 000 coulombs		

This tell us that if we pass 192 000 coulombs through a solution containing Cu^{2+} ions we should get 64 g of copper - the mass of a mole. So, 96 000 coulombs will give us 32 g and so on.

We can do this for any half equation, for example

Al^{3+}	+	$3e^-$	$\rightarrow$	Al
1 mole		3 moles		1 mole
27 g		3 faradays		27 g

We need 3 faradays to produce 27 g of aluminium – that is $3 \times 96\,000$ coulombs. Notice that the charge on the ion tells you how many faradays are needed to produce a mole of element.

7 Fill in the spaces in the table below.

Metal	Ion	Half equation at cathode	No. of faradays needed for 1 mole	1 faraday will produce ...	
				No. of moles	Mass/g
Ca	Ca^{2+}	$Ca^{2+} + 2e^- \rightarrow Ca$	2	$\frac{1}{2}$	20
Na	Na^+		1	1	
Mg					
Al					
Cu	Cu^{2+}				

Working out the amount of charge

You need to know that:

$$\text{charge} = \text{current} \times \text{time}$$
$$Q = I \times t$$
$$\text{coulombs} = \text{amps} \times \text{seconds}$$

Remember the time is always in seconds; 1 minute is 60 seconds, 1 hour is 3600 seconds.

If we know the current (in amps) and the length of time (in seconds) for which it flowed, we can work out the amount of charge in coulombs.

10 amps flowed for 19 200 seconds. How many coulombs is this? How many faradays is this?

$$\text{coulombs} = \text{amps} \times \text{seconds}$$
$$= 10 \times 19\,200 = 192\,000 \text{ coulombs}$$

But 96 000 coulombs = 1 faraday,

so the number of faradays is $\frac{192\,000}{96\,000} = 2$ faradays

8 Explain why 2 faradays produces 40 g of calcium but only 2 g of hydrogen.

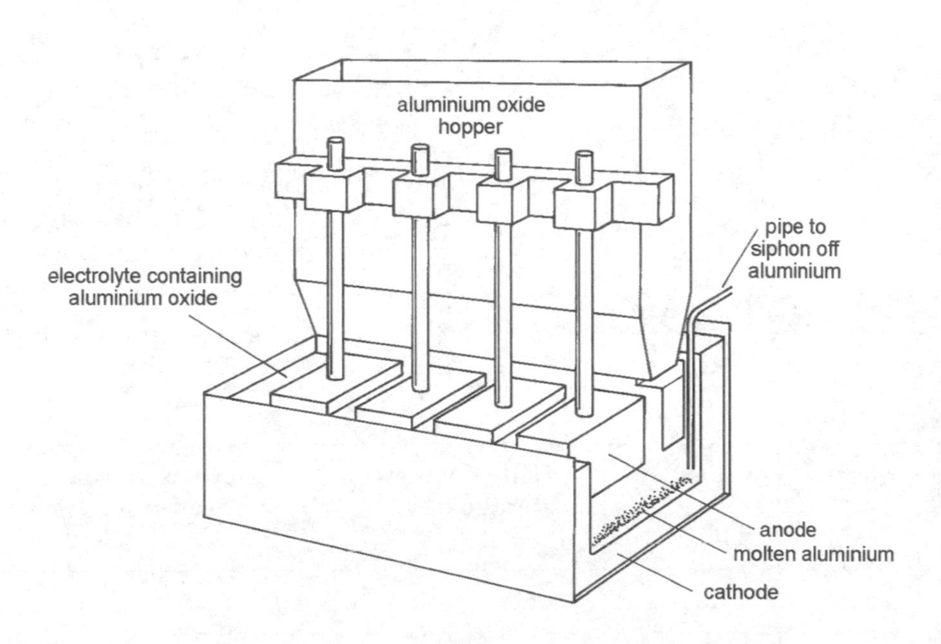

9.9 The industrial production of aluminium.

In the industrial production of aluminium, a current of 30 000 A is passed through molten aluminium oxide for 24 hours. How much aluminium metal is discharged at the cathode?

Answer: First write the half equation and add moles and faradays

Al^{3+}	+	$3e^-$	→	Al
1 mole		3 moles		1 mole
27 g		3 F		27 g

Then work out how many faradays have been passed.

Work out the coulombs first:

$$\text{coulombs} = \text{amps} \times \text{seconds}$$
$$= 30\,000 \times 24 \times 60 \times 60$$
$$= 2\,592\,000\,000 \quad \text{or} \quad 2.592 \times 10^9$$

See Chapter 11 if you are not happy with the use of numbers expressed like 2.592×10^9.

Next work out how many faradays by dividing by 96 000

$$\frac{2\,592\,000\,000}{96\,000} = 27\,000 \text{ F}$$

Now go back to the equation:

3 F produces 27 g of aluminium

so 1 F produces 27/3 g = 9 g

so 27 000 F produces 9 × 27 000 g = 243 000 g = 243 kg

So 243 kg of aluminium is discharged at the cathode.

9 A current of 15 000 amps was passed for 12 hours through molten sodium chloride. How much sodium was produced? Use the following steps: write the equation, work out the number of coulombs, work out the number of faradays.

10 A current of 0.1 amps was passed for 10 minutes through a solution of copper sulphate ($CuSO_4$) using copper electrodes. How much copper was deposited at the cathode? How much copper would be lost at the anode? Use the same steps as in question 9.

Redox

In Chapter 4 you learned that oxidation is loss of electrons and reduction is gain of electrons – OIL RIG. Since metal ions gain electrons at the cathode, they are **reduced** here. Similarly since non-metal ions lose electrons at the anode, **oxidation** always occurs here. One way that you might remember this is to note that cathode and reduction both have a c.

Chapter 10

Physics background

STANDING POINTS

- Do you understand the following terms: **energy**, **electron**? Write down a sentence or two (no more) to explain what you understand by them, then check with the glossary.

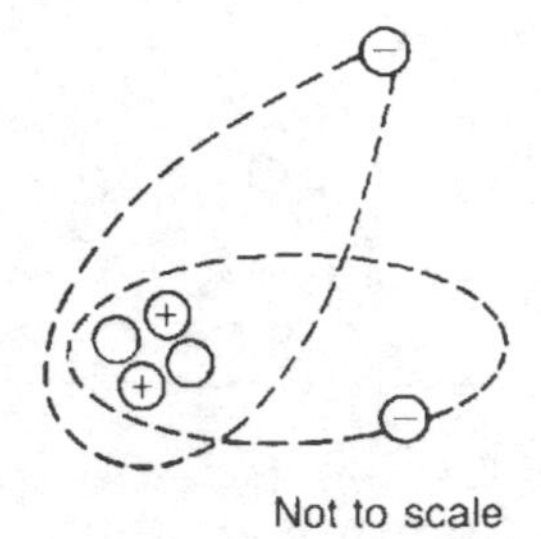

10.1 A helium atom. It is held together by the attraction of opposite electric charges. The two protons in the nucleus are held together against their electrical repulsion by an even stronger force – the nuclear force.

None of the sciences stands alone – to fully understand an advanced course in chemistry, you will need to follow some ideas which come under the heading of physics.

Electricity

Electricity and chemistry are closely linked – it is the attraction of opposite electric charges that holds electrons in atoms.

Electric charge

There are two sorts of electric charge, positive and negative. Like charges repel, so two electrons (each negative) repel and two protons (each positive) also repel. Opposite charges attract so that a proton and an electron attract. Charge is measured in **coulombs**, symbol C. An electron has a charge of 1.6×10^{-19} coulombs so it takes 6.25×10^{18} (625 followed by 16 zeros) electrons to make a coulomb of charge.

Electric circuits

The following symbols are used in electrical circuit diagrams:

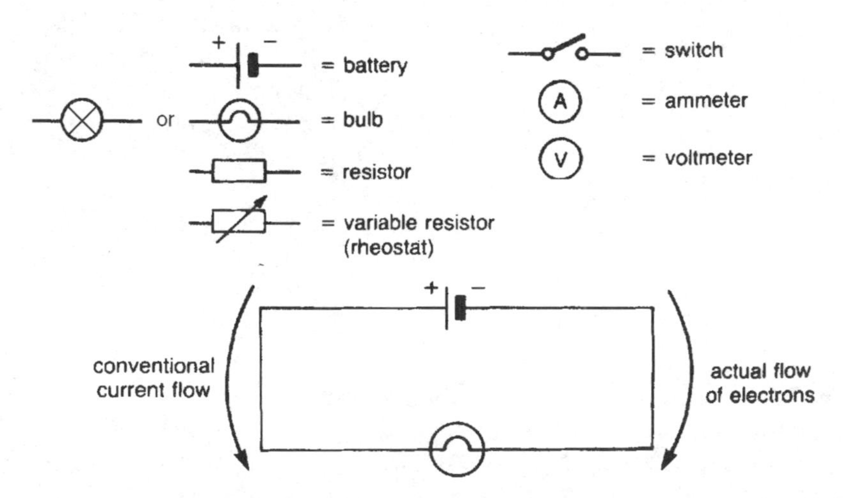

10.2 The directions of current flows in a circuit.

In an electric circuit, charge flows through wires. The actual charge is carried by electrons. These have a negative charge and therefore flow from the negative terminal to the positive. Often we picture imaginary positive charge-carriers flowing from positive to negative. This is called the conventional current. See Fig 10.2.

Electric current

The rate at which electric charge flows through a circuit is called the **current** and is measured in amps (short for amperes). A large current could be produced by a lot of charge moving slowly or a small amount of charge moving fast. A current of one amp is a flow of charge of one coulomb per second.

Amps are coulombs per second.

Using the symbols:

Charge, Q, measured in coulombs
Current, I, measured in amps
Time, t, measured in seconds

$$I = Q/t$$

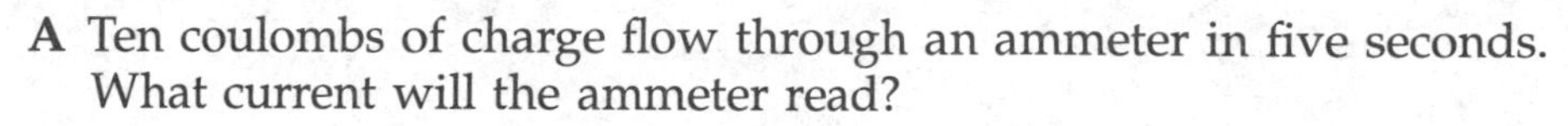

Example

A Ten coulombs of charge flow through an ammeter in five seconds. What current will the ammeter read?

$I = Q/t$

$I = 10/5$

$I = 2$ amps

B A TV set draws a current of 1.6 amps for 10 000 seconds (about 3 hours). How many coulombs of charge pass through the set?

$I = Q/t$, so $Q = I \times t$

$Q = 1.6 \times 10\,000$ C

$Q = 16\,000$ C

Incidentally, since an electron has a charge of 1.6×10^{-19} C, $16\,000/1.6 \times 10^{-19} = 1 \times 10^{23}$ (1 followed by 23 zeros) electrons passed through the TV.

1 What is the current if

a four coulombs flows past a point in two seconds?
b 100 coulombs flows past a point in 20 seconds?
c 360 coulombs flows past a point in a minute?

2 How many coulombs of charge pass a point if

a a current of one amp flows for 10 seconds?
b a current of four amps flows for one hour?

Potential difference

Electric charge flows (called an electric current) if there is a difference in electrical **potential** between two points in a circuit. This is rather like the fact that a ball will roll if there is a height difference between two points. This **potential difference** gives energy to the electric charge. It is measured in volts and is often loosely called **voltage**.

A power supply of one volt gives a charge of one coulomb one **joule** (symbol J) of energy. The charge then transfers this energy into other forms (eg heat, light etc) in the components in the circuit. The charge is not used up, it merely travels around the circuit carrying energy. There is an analogy with postmen delivering letters. The postmen (the charge, measured in coulombs) pick up letters (the energy, measured in joules) at the post office (power supply). They deliver the letters (transfer their energy) as they go round their route and return empty-handed to the post office. The postmen are not used up!

Volts are joules per coulomb.

Using the symbols:

Potential difference, V, measured in volts
Charge, Q, measured in coulombs
Energy, E, measured in joules

$$V = E/Q$$

Example

While starting a car, a 12 V battery supplies a current of 200 A for 5 seconds. How much energy has it supplied?

200 A for 5 seconds means the total charge supplied was $200 \times 5 = 1000$ C.

$V = E/Q$, so $E = Q \times V$
$E = 1000 \times 12$
$E = 12\,000$ J

3 A charge of 2 C is delivered by a 1.5 V battery. How much energy does the battery supply?

Batteries

Given the chance, all metal atoms try to form positive ions by giving away electrons. So in the set up in Fig 10.3, zinc atoms try to form Zn^{2+} ions.

$$Zn \rightarrow Zn^{2+} + 2e^- \text{ (equation 1)}$$

In the same way, copper atoms try to form Cu^{2+} ions.

$$Cu \rightarrow Cu^{2+} + 2e^- \text{ (equation 2)}$$

Zinc is the more reactive, so zinc atoms turn into zinc ions more readily than copper turns into copper ions and more electrons build up on the zinc terminal than on the copper one. They can then flow through the circuit to the copper. This forces copper ions to accept electrons and become copper atoms. This is the reverse of equation 2 above. The net result is that zinc dissolves, copper is deposited and electrons flow through the circuit. The energy for this comes from the chemical reactions. This is the basis of the electric cell called the Daniell cell which produces a potential difference of 1.1 V. Torch batteries work on the same principle.

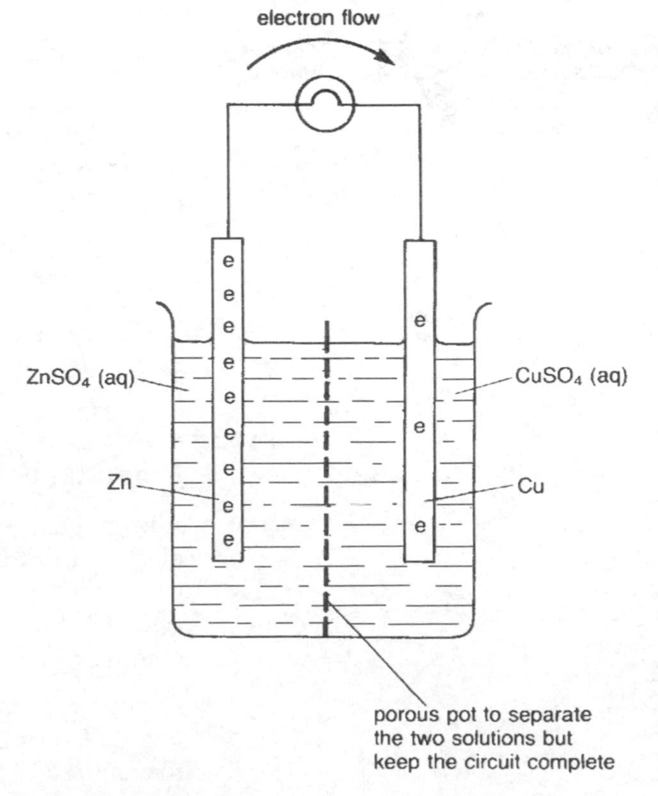

10.3 The Daniell cell.

Resistance

Every component in an electric circuit (e.g. bulb, motor, etc.) has some **resistance** to the flow of electricity. This is measured in **ohms**, (symbol Ω). The bigger the resistance, the smaller the flow of current through the component. But, the bigger the potential difference across the component, the greater the current flow through it. The current flowing through a component can be found using the relationship which is sometimes called **Ohm's law**:

I = V/R

where

I = current in amps
V = potential difference in volts
R = resistance in Ω

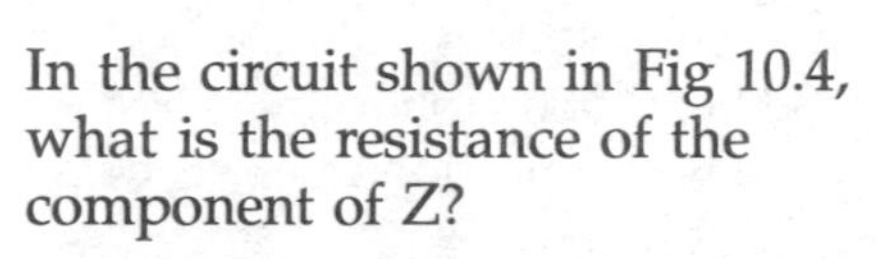

Example

In the circuit shown in Fig 10.4, what is the resistance of the component of Z?

$I = V/R$, so $R = V/I$
$R = 12/2 = 6\ \Omega$

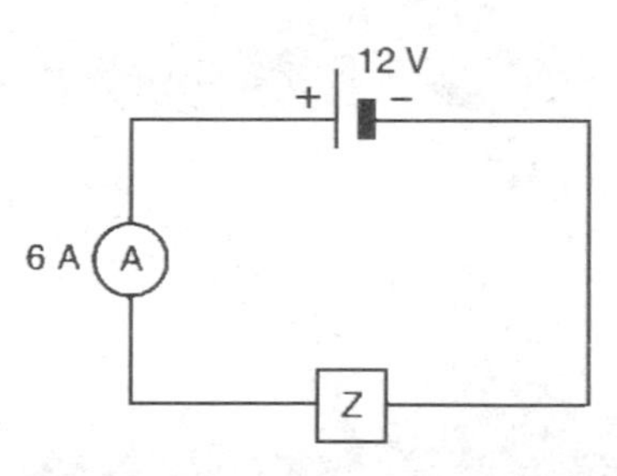

10.4

Exercise

4 What current would flow in the circuit in Fig 10.5, if the values of V and R were:

a $V = 12$ V, $R = 24\ \Omega$
b $V = 240$ V, $R = 480\ \Omega$
c $V = 1.5$ V, $R = 3\ \Omega$?

5 A torch bulb needs a current of 0.1 A and is powered by a 3 V battery. What must its resistance be?

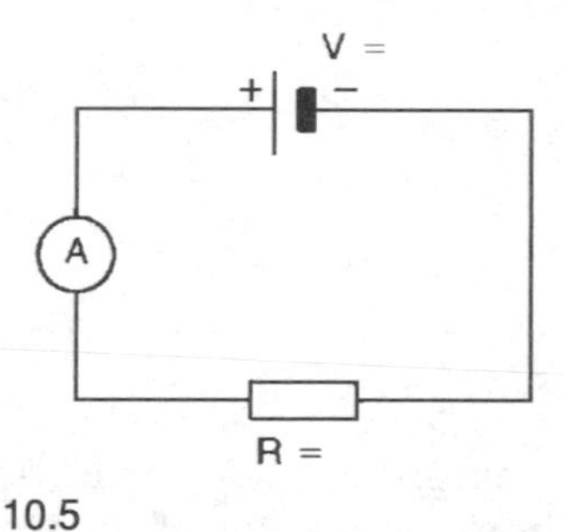

10.5

6 What potential difference do we need to drive a current of 2 A through a resistance of 8 Ω?

Waves

Waves are regular disturbances in a **medium** like ripples in a pond. They carry energy through the medium. For example, throwing a stone into a pond produces a wave through the water which carries energy to move a toy boat up and down. See Fig 10.6.

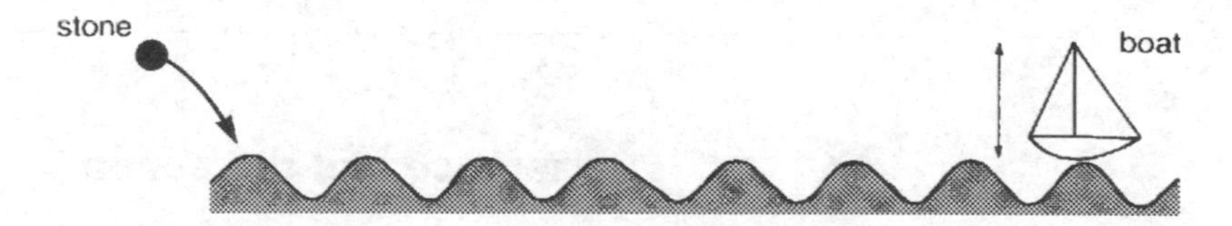

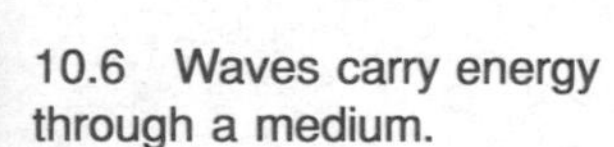
10.6 Waves carry energy through a medium.

A wave has three important properties which we can measure.

1 The **wavelength**, symbol λ (lambda). This is the distance between two successive wave crests (or troughs, or any other identical points on the wave). It is measured in metres (or cm, or other length units).

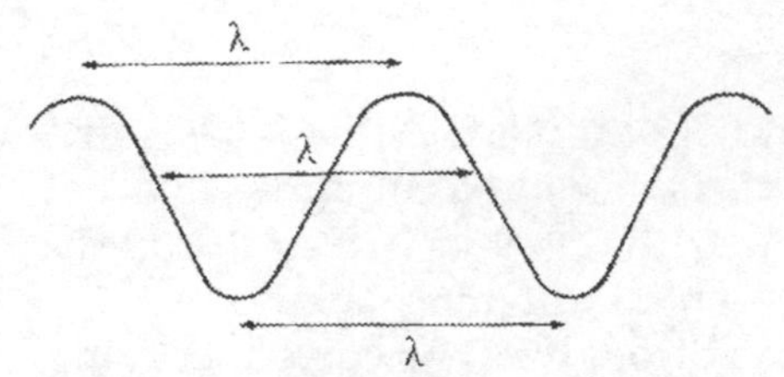

10.7 The wavelength, λ, is the distance between any successive pair of identical points on the wave.

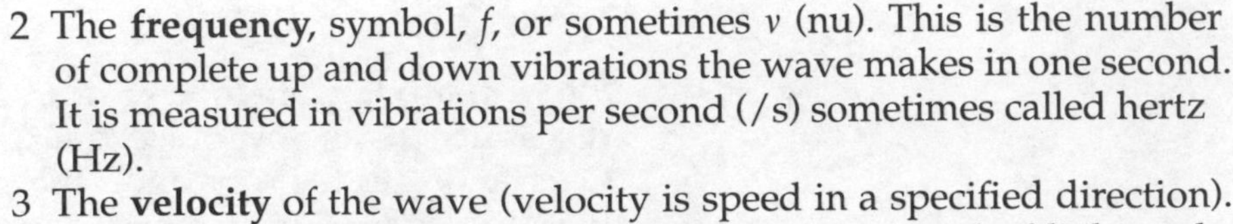

2 The **frequency**, symbol, f, or sometimes ν (nu). This is the number of complete up and down vibrations the wave makes in one second. It is measured in vibrations per second (/s) sometimes called hertz (Hz).

3 The **velocity** of the wave (velocity is speed in a specified direction). This has the symbol v (or sometimes c for the speed of light and other electromagnetic radiation, such as radio wave, X-rays, etc.). This is the distance travelled by the wave in a given time. It is usually measured in metres per second (m/s).

These three quantities are linked by the equation

$v = f\lambda$.

The velocity of light is 300 000 000 m/s. What is the frequency of radio waves of wavelength 1500 m?

$v = f\lambda$, so $f = v/\lambda$
$f = 300\ 000\ 000/1500$
$f = 200\ 000$ vibrations per second (or Hz)

7 A girl produces a wave in a skipping rope by flicking one end up and down twice per second. The wavelength is one metre. What is the speed of the wave along the rope?

8 The frequency of Radio 3 is about 90 000 000 Hz. The speed of all electromagnetic radiation (which includes radio) is 300 000 000 m/s. What is the wavelength of Radio 3?

Using light to probe atoms

You will have been told that electrons exist in 'shells' of different distance from the nucleus of the atom but how do we know this? A lot of our knowledge of electron arrangements comes from studying how light waves are absorbed and emitted by atoms. Light is absorbed and given out by atoms only at particular frequencies, which correspond to different colours. This is why sodium street lights are yellow and neon lights red. Each time a burst of light is absorbed by an atom, it forces an electron to jump up from one shell to the next. When the electron drops back down again, the atom gives out light. By studying the frequency of the light which atoms give out and take in, we can find out about the spacing of their electron shells.

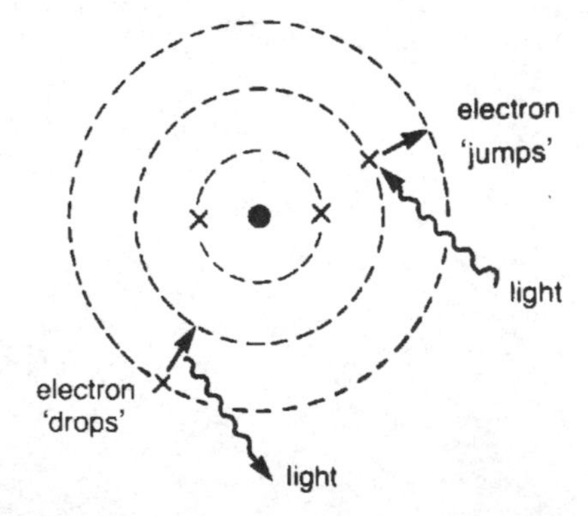

10.8 When electrons jump from shell to shell in an atom, they give out or absorb packets of energy called quanta.

Interference and diffraction

These are two interesting properties of waves. **Diffraction** is shown in the photograph in Fig 10.9.

When waves pass through a small gap (about the same size as their wavelength) they spread out as shown in Fig 10.9 rather than as shown in Fig 10.10, which is what you might expect.

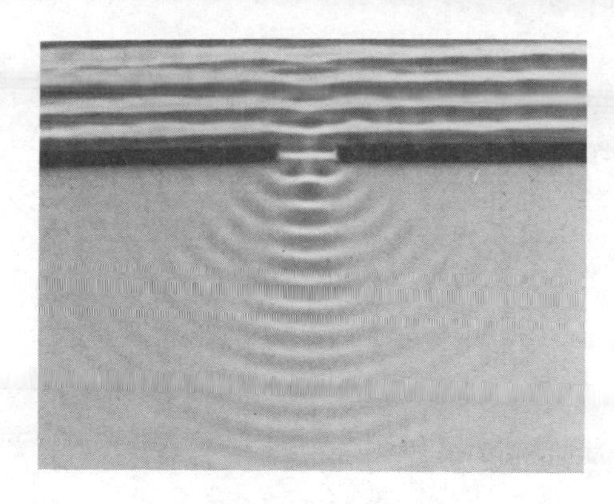

10.9 Diffracted waves spread out after passing through a gap.

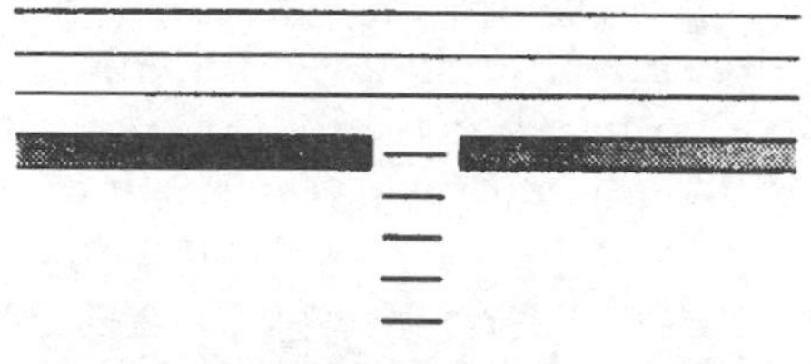

10.10 If waves travelled like bullets they would not spread out after passing through a gap.

Interference occurs when two waves meet. If two crests meet, an extra big crest will be produced. Similarly, two troughs will produce an extra big trough. This is called **constructive interference**. However, a crest meeting a trough will cancel one another out – **destructive interference**.

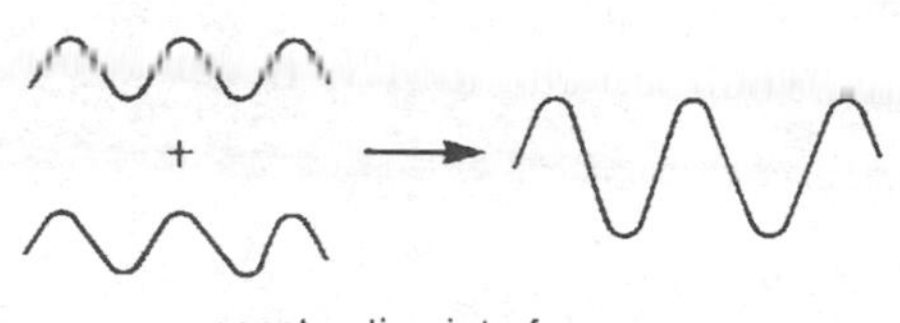

constructive interference

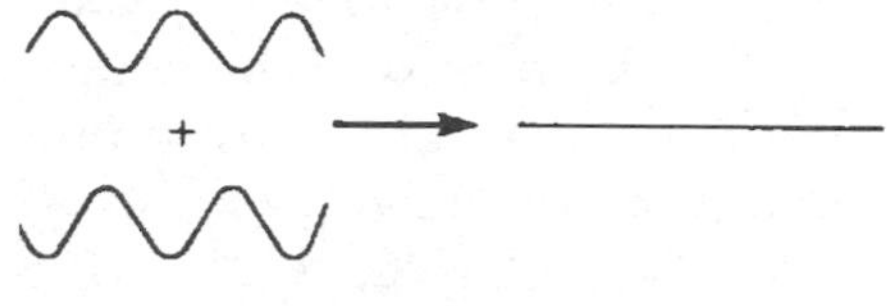

destructive interference

10.11 Waves adding together.

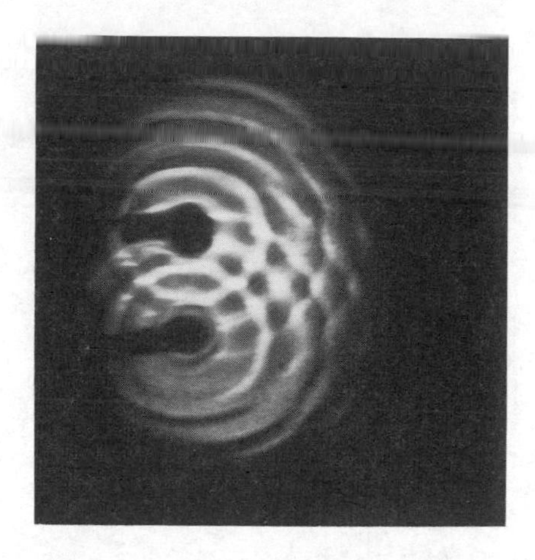

10.12

9 Try to explain the pattern in the photograph above (Fig 10.12) which shows two water waves coming together.

Chapter 11

Mathematics

You don't have to be brilliant at maths to be a good chemist, but certain mathematical skills and techniques are important.

STARTING POINTS

- You will need a scientific calculator – one with the following functions as well as the four basic arithmetical ones:
- square and square root
- log and antilog (to the base 10 and base e)
- exponential
- x^y
- bracket
- sin, cos and tan (ideally)

Equations

In all sciences we look for patterns and if these are very precise we can turn them into mathematical relationships. It might be difficult to fit a mathematical expression to the connection between 'how many fillings in your teeth' and 'how many times you brush your teeth a day' although there could well be a relationship. But we can take more straightforward measurements of things that vary with each other (called **variables**) and write equations which show the connection.

If we double the temperature (in **kelvin**) of a fixed volume of gas, the pressure doubles too. Mathematically speaking, the pressure, *P*, is directly proportional to the temperature, *T*. We write:

$$P \propto T$$

$\propto$ means 'is proportional to'

This is shown by the data in Table 11.1

Table 11.1

Temperature/ K	Pressure/ Pa
100	1000
150	1500
200	2000
250	2500

This also means that the pressure, *P*, is equal to some constant number, *k*, multiplied by the temperature.

$$P = kT$$

In this case, the constant number is 10: if we multiply the temperature in K by 10, we get the pressure in Pa (Pascals).

As the pressure on a gas, which is kept at a constant temperature, goes up, its volume goes down. More precisely, if we double the pressure we halve the volume. Mathematically speaking volume, *V*, is *inversely* proportional to pressure, *P*. Or volume is proportional to $1/P$.

$$V \propto 1/P$$

or $V = k \times 1/P$

or simply $V = k/P$

This is shown by the data in Table 11.2.

Table 11.2

Pressure/ Pa	Volume/ l	1/P/ 1/Pa
1	24	1
2	12	0.5
3	8	0.33
4	6	0.25

1 pascal (Pa) is a unit of pressure of 1 newton per m^2

In this case the constant number is 24. If we multiply $1/P$ by 24, we get the volume.

1 a What would happen to the volume of a gas if we tripled the pressure at constant temperature?
  b If two variables *x* and *y* are directly proportional to each other, what happens to one if we quadruple the other?
  c Write an expression that means *x* is proportional to *y*.

Handling equations

If you can confidently do question 2, then move to the section, **Substituting into equations**, on page 87 and try that. Otherwise go to the section below, **Rearranging equations**.

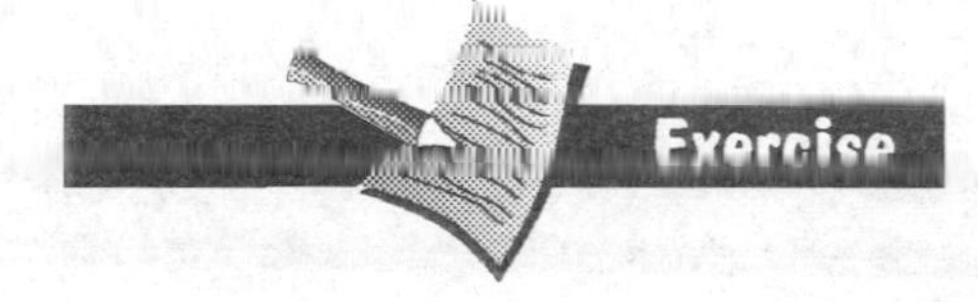

2 The equation that connects the pressure P, volume V and temperature T of a mole of gas is

$$PV = RT$$

where P, V and T are variables and R is a constant called the gas constant.

Rearrange the equation to find

a P in terms of V, R and T
b V in terms of P, R and T
c T in terms of P, V, and R
d R in terms of P, V and T

Rearranging equations

If we are given the expression $a = b/c$ and we want to find a, we substitute the values of b and c into our equation, so that if $b = 10$ and $c = 5$, then $a = 10/5 = 2$. But what do we do if we need to find b or c?

We need to rearrange the equation so that b (or c) appears on its own on the left-hand side of the equation like this:

$$b = ?$$

There are many techniques for doing this. If you are not happy with the method shown below then find an algebra book and revise the method you are more familiar with.

$a = b/c$ can be written $a = \dfrac{b}{c}$ or $a = b \div c$

Step 1: b is divided by c so we multiply both sides by the letter which prevents b being on its own, i.e. c.

So now $c \times a = \dfrac{b \times c}{c}$ usually written $ca = \dfrac{bc}{c}$

This is fair because we have done the same to both sides of the equation.

We can now cancel the c's on the right since b is being both multiplied and divided by c.

$$c \times a = \frac{b \times \not{c}}{\not{c}}$$

Which leaves $c \times a = b$

or $b = c \times a$ usually written in shorthand as $b = ca$.

We can now rearrange this equation in the same way to find c.

A handy rule

For simple expressions of the form $a = \frac{b}{c}$ $x = \frac{y}{z}$ etc.

some people memorise a triangle

$$\frac{b}{a \times c} \qquad \frac{y}{x \times z}$$

From this they can see directly, by covering up the one they want that:

$$a = \frac{b}{c}; \quad x = \frac{y}{z}$$

$$c = \frac{b}{a}; \quad z = \frac{y}{x}$$

$$b = a \times c \quad y = x \times z$$

Step 2: $b = c \times a$. Divide both sides by a.

$$\frac{b}{a} = \frac{c \times a}{a}$$

We can now cancel the a's on the right.

$$\frac{b}{a} = \frac{c \times \not{a}}{\not{a}}$$

So $b/a = c$ or $c = b/a$

Notice that because c started on the bottom, a two-step process was necessary. We found an expression for b first and then found one for c.

3 Try the following examples to test your understanding of the method, finding the variable in brackets in terms of the others.

a $p = \frac{q}{r}$ (q) b $n = mt$ (m)

c $g = \frac{fe}{h}$ (h) d $\frac{pr}{oq} = s$ (r)

Cross multiplying

A useful short cut follows from the method above for rearranging equations.

Take the equation: $ab = cd$

Any variable can be moved to the other side of the equation, but if it was multiplying on one side it must divide on the other, and if it was dividing on one side it must multiply on the other.

So $a = \frac{cd}{b}$ $b = \frac{cd}{a}$

If you think of any equation as having the shape below then you can change the position of the variables by following the arrows:

$$\frac{a}{c} \rightleftarrows \frac{d}{b}$$

If one or more of a, b, c and d are not single variables then move everything.

For example $(a+b) = \frac{d}{c}$

the whole bracket $(a+b)$ must be moved if you change it to the other side.

$$c = \frac{d}{(a+b)}$$

A pitfall to avoid

Calculators are great but you must use them with care. Here is a common mistake to watch out for.

Imagine you are working out the expression $z = \dfrac{24}{3 \times 4}$

This means that z is 24 divided by (3×4) which is 12, so that $z = 2$.

But it is easy to punch into your calculator 24 ÷ 3 × 4 = which gives the answer 32 which is incorrect.

What you must do is punch in 24 ÷ 3 ÷ 4 = which gives the correct answer of 2.

Alternatively you could use brackets and punch in 24 ÷ (3 × 4) = which also gives the answer 2.

'BIDMAS'
A useful way to remember the order in which to tackle the operations in mathematical equations is summarised with the word **BIDMAS.**

- **B**rackets
- **I**ndices
- **D**ivision
- **M**ultiplication
- **A**ddition
- **S**ubtraction

Substituting into equations

You will often be given an expression in the form of letters and you will have to substitute the correct value for each letter before you can carry out the calculation.

You must then carry out the various mathematical operations in the correct order in your head or using a calculator.

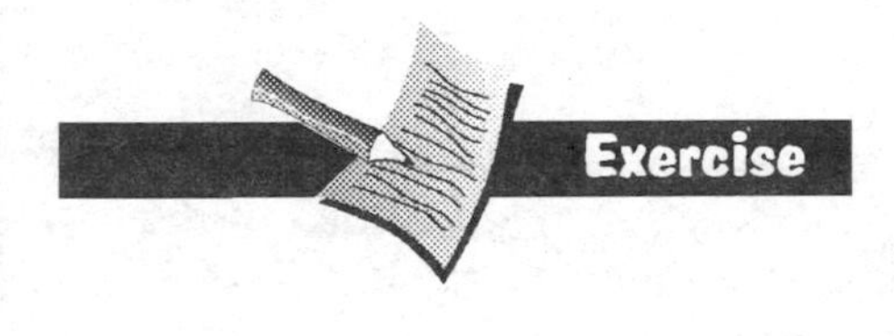

4 Find the value of x in the expressions below using your calculator.

$a = 0.1, b = 17, c = 14$

a $x = \dfrac{2ab}{c}$ b $x = \dfrac{a}{bc}$

c $x = \dfrac{(a-b)(c-a)}{2b}$ d $x = a^2(b-a)$

Some formulae
Expressions you will regularly encounter in advanced chemistry calculations for which the practice you have just done is very useful are:

$\text{Density} = \dfrac{\text{Mass}}{\text{Volume}}$

$\text{No. of moles} = \dfrac{\text{Mass}}{\text{Mass of 1 mole}}$

$\text{No. of moles in a solution} = \dfrac{\text{Molarity} \times \text{Volume}}{1000}$

Powers

If we multiply a number by itself we say we have squared it, so 3×3 is three squared, written 3^2.

$3 \times 3 \times 3$ is three cubed, written 3^3.

$3 \times 3 \times 3 \times 3$ is called three raised to the power 4, written 3^4.

$3 \times 3 \times 3 \times 3 \times 3$ is called three raised to the power 5 written 3^5 and so on.

Using this notation, a number raised to the power 1 is just the number itself, so that, for example, $5^1 = 5$.

Any number raised to the power 0 is 1, so that, for example, $5^0 = 1$.

Your calculator will probably have a 'squaring' button (usually marked x^2). To raise a number to any other power, use the x^y button as in the following example.

To find 3^4, punch into your calculator 3 x^y 4 = and you should get the answer 81.

5 Find the following using your calculator.

a 8^2 b 2^8 c 3^3

Numbers expressed as powers of ten

It is important that you understand the meaning of numbers expressed using powers of ten in the way shown below.

1×10^2 means $10 \times 10 = 100$
1×10^3 means $10 \times 10 \times 10 = 1000$
1×10^4 means $10 \times 10 \times 10 \times 10 = 10\ 000$
2×10^4 means $2 \times 10 \times 10 \times 10 \times 10 = 20\ 000$

This is an important shorthand for extremely large numbers. It can be extended to small numbers too.

1×10^{-2} means $\dfrac{1}{10 \times 10} = 0.01$

1×10^{-3} means $\dfrac{1}{10 \times 10 \times 10} = 0.001$

1×10^{-4} means $\dfrac{1}{10 \times 10 \times 10 \times 10} = 0.0001$

2×10^{-4} means $\dfrac{2}{10 \times 10 \times 10 \times 10} = 0.0002$

As these numbers get very big or very small, the beauty of this shorthand will become apparent. You will often need to use numbers such as 10^7 or 10^{-7} so think how tedious it is to write 10 000 000 or 0.000 000 1 at every stage of a calculation.

6 Write the following numbers in two other ways as above.

a 1×10^7 b 100 000
c $3 \times 10 \times 10 \times 10 \times 10 \times 10$ d 1×10^{-7}
e 0.000 000 001 f $\dfrac{1}{(10 \times 10 \times 10 \times 10 \times 10 \times 10)}$

To multiply numbers expressed like this we *add* the powers (called indices) and to divide we *subtract* them.

For example: $1 \times 10^5 \times 1 \times 10^3 = 1 \times 10^8$

We can check this by writing the numbers in full:
$100\ 000 \times 1000 = 100\ 000\ 000$

and $1 \times 10^5 / 1 \times 10^3 = 1 \times 10^2$

Check this yourself by writing the numbers in full.

7 Calculate the following in your head.

a $1 \times 10^4 \times 1 \times 10^4$
b $1 \times 10^4 / 1 \times 10^3$
c $1 \times 10^5 / 1 \times 10^7$

Very big and very small numbers

Some numbers you will often meet in chemistry are:

The Avogadro constant (the number of particles in a mole): 6×10^{23}

The mass of a hydrogen atom: 1.6×10^{-27} kg

The charge of the electron: -1.6×10^{-19} C

The speed of light: 3×10^{8} m/s

Imagine writing these in full each time!

Numbers like 1×10^{3} and 1×10^{-7} are often written 10^{3} and 10^{-7} respectively. You need to remember the 1 when punching them into a calculator.

The EXP button on your calculator

It is vital that you can use your calculator correctly and efficiently to work with numbers like this. The EXP button stands for 'times 10 raised to the power of'.

To use your calculator to type in 1×10^{3}, press

1 EXP 3 and you will see displayed 1. 03 (or $1.^{03}$ on some calculators). Don't confuse this with 1.03.

Similarly to put 1×10^{-3} into the calculator, press

1 EXP 3 +/− and you will see displayed 1.–03 (or $1.^{-03}$ on some calculators).

Use the +/− button on your calculator for the sign of powers of 10, not the subtraction button.

Make sure you are familiar with the type of display your calculator gives.

When you write down an answer from a calculator display, remember to write the number in full so that a display of 1. 03 (or $1.^{03}$ is written 1×10^{3} and a display of 1.–03 (or $1.^{-03}$) as 1×10^{-3}.

8 Try the following calculations, using your calculator. It is a good idea to estimate your answer to see if you are using the calculator correctly.

a $\dfrac{(1 \times 10^{3}) \times 14}{6}$ b $(1 \times 10^{3}) \times (3 \times 10^{9})$

c $\dfrac{6 \times 10^{23}}{1.38 \times 10^{-23}}$

More numbers

The following are all equivalent.

1000	
1×10^{3}	0.1×10^{4}
10×10^{2}	0.01×10^{5}
100×10	0.001×10^{6}
1000×10^{0}	0.0001×10^{7}

but the most usual way of writing this is 1×10^{3}. This is called **standard form**.

Using logarithms

It is probably more important to a chemist that he or she can *use* logarithms than that they fully remember the origins of this mathematical idea – but for the curious, the concept of a logarithm or **log** is briefly explained below.

The $\log_{10}$, which is read 'log to the base 10' of a number is the value of the power to which 10 is raised to give the number.

$1 = 1 \times 10^0$ so $\log_{10} 1 = 0$
$10 = 1 \times 10^1$ so $\log_{10} 10 = 1$
$100 = 1 \times 10^2$ so $\log_{10} 100 = 2$
$1000 = 1 \times 10^3$ so $\log_{10} 1000 = 3$
$10000 = 1 \times 10^4$ so $\log_{10} 10000 = 4$

This should show you that using logs is a convenient way of scaling down very large numbers.

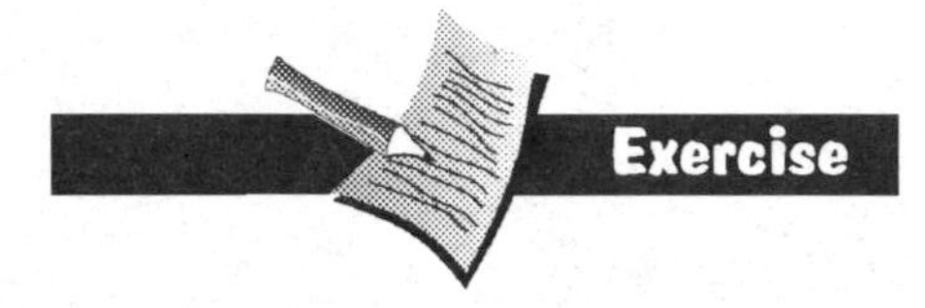

9 Work out $\log_{10}$ of 1 000 000.

The pH scale

	pH	concentration of H^+/ mol/l
'bench' hydrochloric acid →	0	1.0 = 1×100
	1	0.1 = 1×10^{-1}
stomach acid →	2	0.01 = 1×10^{-2}
cola →	3	0.001 = 1×10^{-3}
	4	0.0001 = 1×10^{-4}
	5	= 1×10^{-5}
rain →	6	= 1×10^{-6}
tap water →	7	= 1×10^{-7}
	8	= 1×10^{-8}
	9	= 1×10^{-9}
some household → cleaners	10	= 1×10^{-10}
	11	= 1×10^{-11}
	12	= 1×10^{-12}
lime water →	13	= 1×10^{-13}
'bench' sodium hydroxide →	14	= 1×10^{-14}

You will certainly have come across this scale of measuring acidity and alkalinity, but why does it use such odd numbers? Acidity is caused by hydrogen ions, H^+, dissolved in water. The more concentrated the H^+ ions, the more acidic the solution. The concentration of H^+ in 1 mol/l hydrochloric acid ('bench HCl') is 1 mol/l, in pure water it is 10^{-7} mol/l and in 1 mol/l sodium hydroxide ('bench NaOH') it is 10^{-14} mol/l. Yes even in an alkali, there are a few H^+ ions! These numbers cover an inconveniently large range so we get round this by using logs.

$\log_{10} 1 = 0$

$\log_{10} 1 \times 10^{-7} = -7$

$\log_{10} 1 \times 10^{-14} = -14$

This means that in most everyday solutions, the $\log_{10}$ of the H^+ concentration will be a negative number. So we agree to multiply them all by -1 to make them all positive. This is called the **pH**.

so

1 mol/l HCl has pH 0

Pure water has pH 7

1 mol/l NaOH has pH 14

Table 11.3 The pH scale. Could you complete the table?

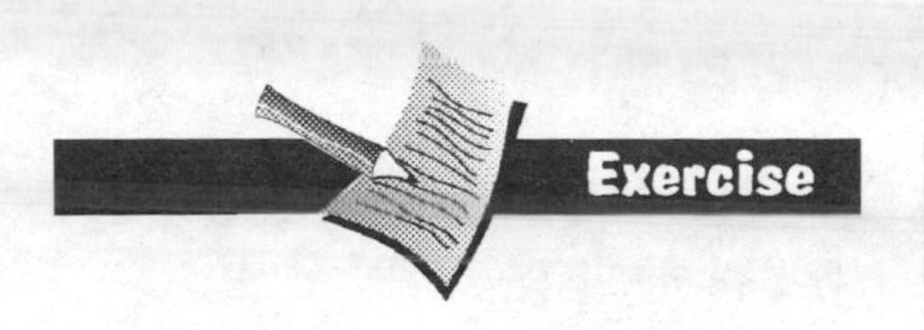

10 Vinegar is a solution of ethanoic acid where the concentration of H^+ is about 10^{-4} mol/l. What is its pH?

The pHs of vinegar and cola are about the same.

Logs, antilogs and your calculator

It is not too hard to work out $\log_{10}$ of numbers like 100 or 1×10^5 but we can find $\log_{10}$ of *any* number using a calculator.

To find $\log_{10} 9$ follow this procedure:

Push the buttons in the order

9 log   and you should see display 0.954242509

11 Try the following examples to see that you can do this correctly. Show your answer to three decimal places.

a $\log_{10} 14$ b $\log_{10} 7$ c $\log_{10} 9.9$
d $\log_{10} 0.7$ e $\log_{10} 0.003$

The answers to (d) and (e) will be negative, since $\log_{10} 1 = 0$, and any number smaller than 1 has a negative $\log_{10}$.

It is also important to be able to reverse the process. Suppose you have the value of a $\log_{10}$ and you want to find the number it came from.

3.5 is the log of what number?

Push the buttons on your calculator in this order

3.5 INV log

You should see display 3162.27766

This is called the **antilog** of 3.5.

so $\log_{10} 3162.27766 = 3.5$

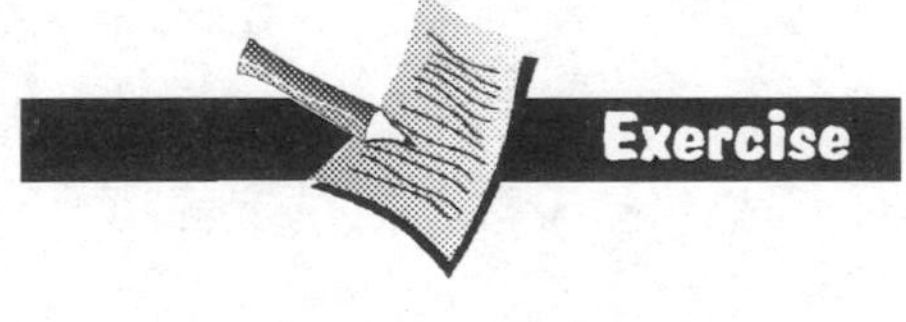

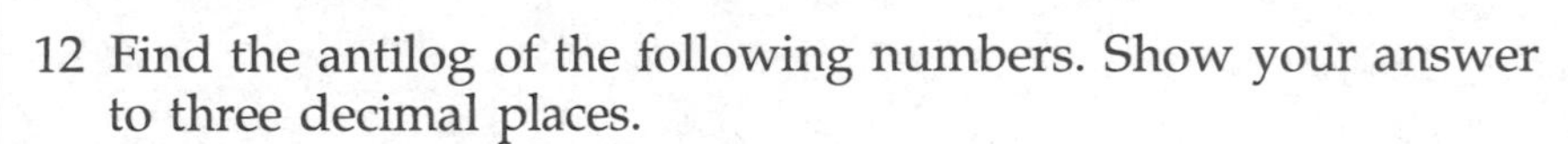

12 Find the antilog of the following numbers. Show your answer to three decimal places.

a 1.7   b 2.8   c 3.9   d 0.2   e 0.07   f 0.0094

Using proportion

There are a lot of proportion calculations in chemistry and there are many ways of approaching them. The following method works perfectly well, but if you have a different method which works for you, then stick with it.

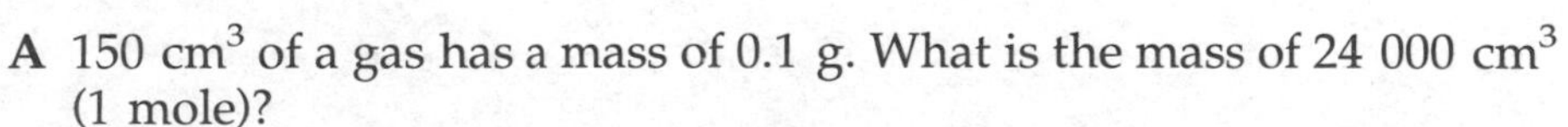

A 150 cm^3 of a gas has a mass of 0.1 g. What is the mass of 24 000 cm^3 (1 mole)?

150 cm^3 of a gas has a mass of 0.1 g

1 cm^3 would have a mass of 0.1/150 g ($= 6.6 \times 10^{-4}$ g)

24 000 cm^3 would have a mass of $6.6 \times 10^{-4} \times 24\,000$ g (=16 g)

This is the mass of one mole of the gas so its M_r must be 16.

B 5.187 g of potassium combines with 2.128 g of sulphur. How many grams of sulphur combine with 39 g (1 mole) of potassium?

5.187 g of potassium combines with 2.128 g of sulphur

1 g of potassium combines with 2.128/5.187 g of sulphur (=0.410 g)

39 g of potassium combines with 0.410×39 g of sulphur (=16 g)

16 g of sulphur is 0.5 mol. So 1 mole of potassium (K) combines with 0.5 mol of sulphur (S) so the formula of the compound must be K_2S.

13 a If 0.125 g of a solid reacts with an acid to produce 10 cm^3 of a gas, how much solid would produce 22 400 cm^3 of gas?
b If 0.0015 g of magnesium reacts with 0.002 g of sulphur, how many grams of sulphur will 24 g of magnesium react with?

Working with graphs

These graphs show the rate of the reaction between magnesium and hydrochloric acid to produce hydrogen. Some mathematical techniques can help us to interpret them.

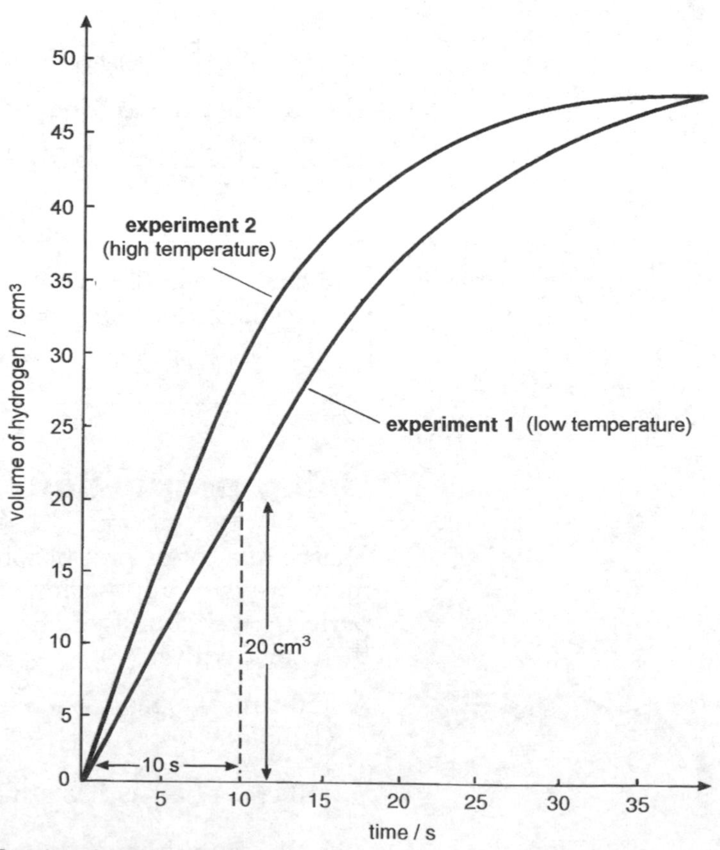

11.1 Graphs of reaction rate

First, the **gradient** of a straight line is found by dividing the length of line A (vertical) by the length of line B (horizontal) (see Fig 11.2).

$$\text{gradient} = \frac{A}{B}$$

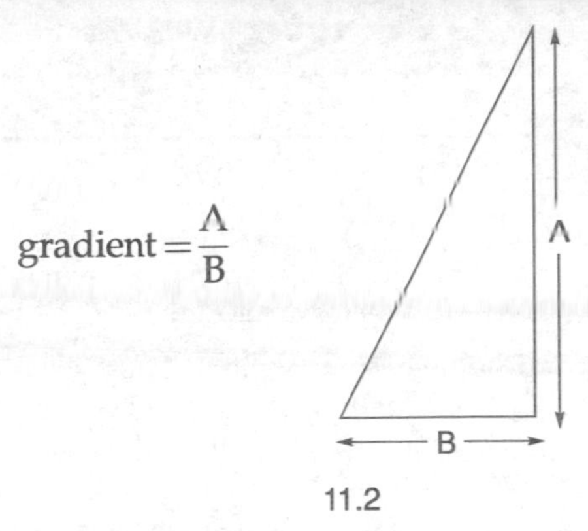

11.2

In the case of experiment 1 above this tells us the rate at which hydrogen is produced, between 0 and 10 seconds. It will have units.

$$\text{rate} = \frac{20\ \text{cm}^3}{10\ \text{s}} = 2\ \text{cm}^3/\text{s}$$

It follows that the steeper the line, the greater the rate.

14 Find the rate of the reaction for experiment 2 over the first 10 seconds.

When we get to the part of the graph where the line starts to curve, the best we can do is to take a tangent at the point we are interested in and find the gradient of the tangent. A tangent is a line drawn so that it just touches the graph line at one particular point.

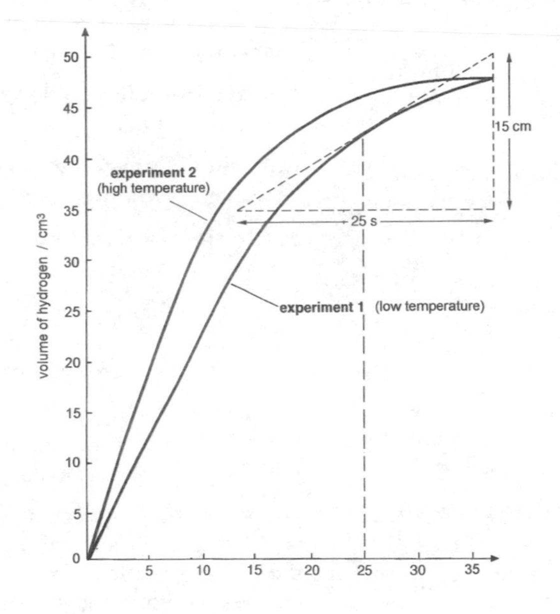

11.3

The rate in experiment 1 after 30 seconds is $\frac{15\ \text{cm}^3}{25\ \text{s}} = 0.6\ \text{cm}^3/\text{s}$

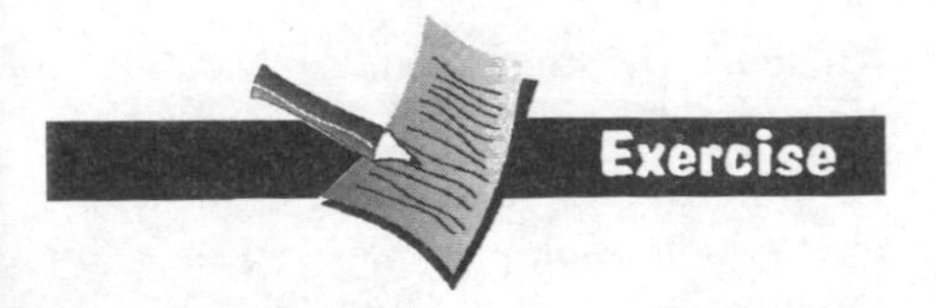

15 Find the rate after 30 s in experiment 2.

Glossary

Activation energy: The minimum energy needed by the reactants before reaction can occur.

Alkane: A hydrocarbon containing only C–C and C–H single bonds.

Alkene: A hydrocarbon with one or more C=C bonds.

Anode: The positive electrode in an electrolysis cell.

Atom: The smallest particle of an element.

Atomic mass: The mass of an atom measured in u. 1 u is 1.6×10^{-27} kg.

Atomic number: The number of protons in the nucleus of an atom – also the order of an element in the Periodic Table.

Balanced equation: An equation where the total number of each type of atom is the same on each side of the arrow.

Boyle's Law: The gas law stating that the product of pressure and volume is constant.

Cathode: The negative electrode in an electrolysis cell.

Charles' Law: The gas law stating that volume is proportional to temperature (measured in kelvin).

Compound: A substance composed of two or more elements chemically bonded together.

Concentration: The amount of substance dissolved in a given amount of solution – usually measured in moles per litre (mol/l).

Constant Volume Law: The gas law stating that pressure is proportional to temperature (measured in kelvin).

Coulomb: The unit of electric charge.

Covalent bonding: Bonding in which the atoms are held together by sharing electrons.

Current: The rate of flow of electric charge – measured in coulombs per second, or amps.

Density: The mass of a given volume of a substance – usually measured in grams per cm^3 (g/cm^3).

Diffraction: The spreading out of waves as they pass through a gap comparable in size with their wavelength.

Double decomposition: A reaction between two soluble salts where the ions 'swap partners' and an insoluble salt is formed.

Electrode: The terminal by which electric current enters or leaves an electrolysis cell (see **anode** and **cathode**).

Electrolysis: The process by which electricity flows through an ionic compound (molten or in solution) and breaks up the compound.

Electrolyte: A liquid containing ions.

Electron: A sub-atomic particle of relative atomic mass 1/1840 and one unit of negative charge. Electrons are found orbiting the nuclei of all atoms. They are found in shells holding successively 2, 8, 8 ... electrons.

Element: One of the 109 substances that cannot be broken down chemically into anything simpler.

Endothermic: This describes a reaction in which heat is taken in from the surroundings to the reaction mixture.

Energy: The ability to do work – measured in joules.

Enthalpy: Heat energy measured under conditions of constant pressure.

Entity: The simplest formula unit of an element or compound.

Exothermic: This describes a reaction in which heat is given out from the reaction mixture to the surroundings.

Frequency: The number of vibrations per second made by a wave – measured in vibrations per second (Hz).

Giant structure: A regular 3-D arrangement of atoms or ions in which the bonding (ionic, covalent or metallic) extends throughout the substance.

Group: A vertical column of elements in the Periodic Table in which all the elements have some similarities because they have the same number of electrons in their outer shells.

Half-equation: An equation for a reaction which takes place at an electrode showing ions gaining or losing electrons.

Interference: The interaction of two waves so that two crests (or two troughs) add together while a trough and a crest cancel out.

Ion: An atom (or group of atoms) having a charge (negative or positive). Ions are formed from atoms by loss or gain of electrons.

Ionic bonding: Bonding in which the atoms are held together by the transfer of electrons from metal to non-metal atoms.

Ionisation energy: The amount of energy needed to remove an electron from an atom.

Isotopes: Two or more varieties of the same element having the same number of protons but different numbers of neutrons (and therefore different relative atomic masses).

Kelvin scale: A temperature scale starting from absolute zero (–273 °C) using degrees the same size as °C. The units are written K (not °K).

Metallic bonding: Bonding in which metal atoms are held together by pooling of electrons.

Metalloid: An element which appears close to the 'staircase line' in the Periodic Table and has some metallic and some non-metallic properties. Also called a semi-metal.

Mole: The relative atomic or molecular mass of a substance in grams. One mole of any substance contains the same number (6×10^{23}) of entities.

Molecular mass: The mass of a molecule measured in u. 1 u is 1.6×10^{-27} kg.

Molecular structure: An arrangement of atoms covalently bonded together in small groups (molecules) which are independent of each other.

Molecule: A small group of atoms held together by covalent bonding.

Neutron: A sub-atomic particle of relative atomic mass 1 and no charge. Neutrons are found in the nuclei of all atoms except hydrogen.

Nucleus: The tiny central core of the atom containing protons and neutrons.

Period: A horizontal row of elements in the Periodic Table.

Potential difference: The amount of energy needed to move one coulomb of charge between two points in a circuit. More loosely the force pushing charge around the circuit. Measured in joules per coulomb or volts. Often called 'voltage'.

Precipitate: Small particles of insoluble solid which are produced in a liquid by a chemical reaction. They sink to the bottom of the liquid.

Proton: A sub-atomic particle of relative atomic mass 1 and with one unit of positive charge. Protons are found in the nuclei of all atoms.

Relative atomic mass: The mass of one atom of an element on a scale on which the mass of a hydrogen atom is one (more precisely, the mass of an atom of ^{12}C is 12).

Relative molecular mass: The mass of a molecule on the relative atomic mass scale.

Resistance: A measure of a component's opposition to the flow of electric current – measured in ohms (Ω).

Salt: An ionic compound formed by the reaction of an acid with a base.

Standard form: A way of writing a number using powers of ten so that there is one digit before the decimal point. For example 4.5×10^3.

Transition element: One of the block of typical metal elements in the centre of the Periodic Table.

Valency: The combining power of an atom.

Wavelength: The distance between two crests, troughs or any other similar points on a wave.

The Periodic Table

Key: Relative atomic mass / Symbol / Name / Atomic number

Period	I (s-block)	II (s-block)	d-block										III (p-block)	IV (p-block)	V (p-block)	VI (p-block)	VII (p-block)	0 (p-block)
1	1.0 H Hydrogen 1																	4.0 He Helium 2
2	6.9 Li Lithium 3	9.0 Be Beryllium 4											10.8 B Boron 5	12.0 C Carbon 6	14.0 N Nitrogen 7	16.0 O Oxygen 8	19.0 F Fluorine 9	20.2 Ne Neon 10
3	23.0 Na Sodium 11	24.3 Mg Magnesium 12											27.0 Al Aluminium 13	28.1 Si Silicon 14	31.0 P Phosphorus 15	32.1 S Sulphur 16	35.5 Cl Chlorine 17	39.9 Ar Argon 18
4	39.1 K Potassium 19	40.1 Ca Calcium 20	45.0 Sc Scandium 21	47.9 Ti Titanium 22	50.9 V Vanadium 23	52.0 Cr Chromium 24	54.9 Mn Manganese 25	55.9 Fe Iron 26	58.9 Co Cobalt 27	58.7 Ni Nickel 28	63.5 Cu Copper 29	65.4 Zn Zinc 30	69.7 Ga Gallium 31	72.6 Ge Germanium 32	74.9 As Arsenic 33	79.0 Se Selenium 34	79.9 Br Bromine 35	83.8 Kr Krypton 36
5	85.5 Rb Rubidium 37	87.6 Sr Strontium 38	88.9 Y Yttrium 39	91.2 Zr Zirconium 40	92.9 Nb Niobium 41	95.9 Mo Molybdenum 42	(99) Tc Technetium 43	101.1 Ru Ruthenium 44	102.9 Rh Rhodium 45	106.4 Pd Palladium 46	107.9 Ag Silver 47	112.4 Cd Cadmium 48	114.8 In Indium 49	118.7 Sn Tin 50	121.8 Sb Antimony 51	127.6 Te Tellurium 52	126.9 I Iodine 53	131.3 Xe Xenon 54
6	132.9 Cs Caesium 55	137.3 Ba Barium 56	138.9 La * Lanthanum 57	178.5 Hf Hafnium 72	181.0 Ta Tantalum 73	183.9 W Tungsten 74	186.2 Re Rhenium 75	190.2 Os Osmium 76	192.2 Ir Iridium 77	195.1 Pt Platinum 78	197.0 Au Gold 79	200.6 Hg Mercury 80	204.4 Tl Thallium 81	207.2 Pb Lead 82	209.0 Bi Bismuth 83	(210) Po Polonium 84	(210) At Astatine 85	(222) Rn Radon 86
7	(223) Fr Francium 87	(226) Ra Radium 88	(227) Ac † Actinium 89	(261) Db Dubnium 104	(262) Jl Joliotium 105	(263) Rf Rutherfordium 106	(262) Bh Bohrium 107	(?) Hn Hahnium 108	(?) Mt Meitnerium 109									

f-block

*Lanthanides	140.1 Ce Cerium 58	140.9 Pr Praseo-dymium 59	144.2 Nd Neodymium 60	(147) Pm Promethium 61	150.4 Sm Samarium 62	152.0 Eu Europium 63	157.3 Gd Gadolinium 64	158.9 Tb Terbium 65	162.5 Dy Dysprosium 66	164.9 Ho Holmium 67	167.3 Er Erbium 68	168.9 Tm Thulium 69	173.0 Yb Ytterbium 70	175.0 Lu Lutetium 71
†Actinides	232.0 Th Thorium 90	(231) Pa Protactinium 91	238.1 U Uranium 92	(237) Np Neptunium 93	(242) Pu Plutonium 94	(243) Am Americium 95	(247) Cm Curium 96	(245) Bk Berkelium 97	(251) Cf Californium 98	(254) Es Einsteinium 99	(253) Fm Fermium 100	(256) Md Mendele-vium 101	(254) No Nobelium 102	(257) Lr Lawrencium 103

Answers

Chapter 1

1 The particles gain extra energy which makes them vibrate more thus pushing one another further apart.

2 The pressure will increase (double). The particles do not change in speed so they hit the walls of the syringe just as hard as before. As the volume is now smaller, they hit the walls more often.

3 a 373 K; 273 J; 773 K
b 227 °C; 200 °C; –73 °C

4 22.0 cm^3

5 a The particles of reactants will collide with one another harder and more often.
b The particles of the other reactant will find more exposed particles of the solid reactant to collide with.
c The reactant particles will collide more often as there is less space in which to move.
d More particles of reactant in the same space will lead to more collisions.

Chapter 2

1

	p	n	e
C	6	6	6
Li	3	4	3
U	92	146	92
Mn	25	30	25

2 a Hydrogen atoms have no neutron.
b Chlorine appears to have 18.5 neutrons.

3

		p	n	e
a	$^{12}_{6}C$	6	6	6
	$^{13}_{6}C$	6	7	6
	$^{14}_{6}C$	6	8	6
b	$^{1}_{1}H$	1	0	1
	$^{2}_{1}H$	1	1	1
	$^{3}_{1}H$	1	2	1

4 20.2

5 a 2,3; b 2,7; c 2,1

6 Only the number of electrons changes. The nucleus, which gives the atom its identity, does not vary.

7 a Ca^{2+}, b O^{2-}, c Al^{3+}, d N^{3-}

8 a

Na
Cl
Na^+
Cl^-

b

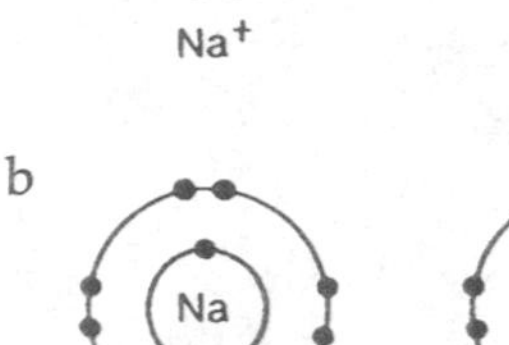

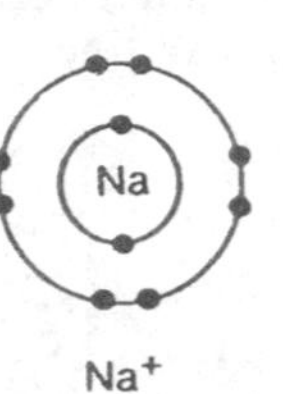

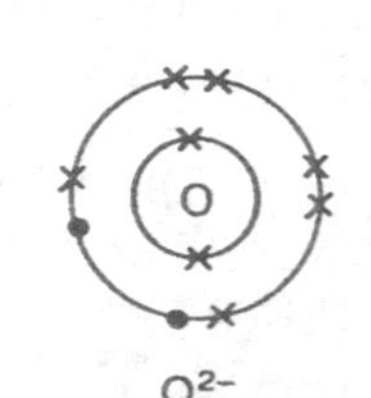

c

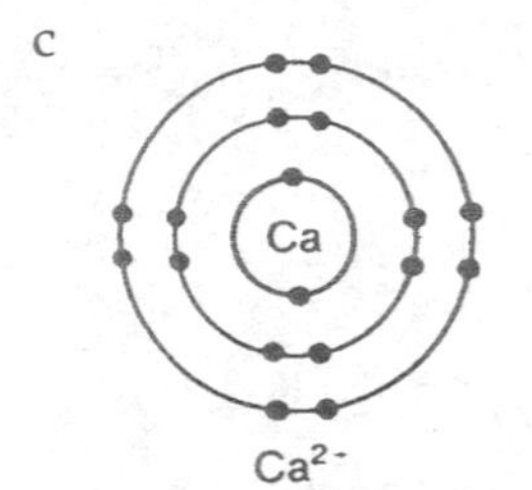

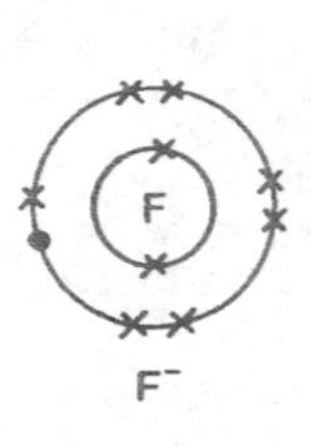

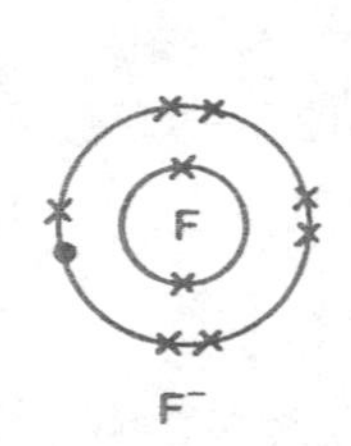

9 a NaCl, sodium chloride; b Na_2O, sodium oxide; c CaF_2, calcium fluoride

10 a

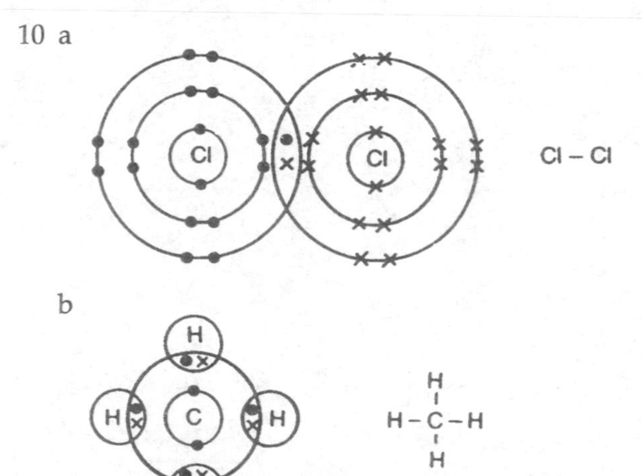

c

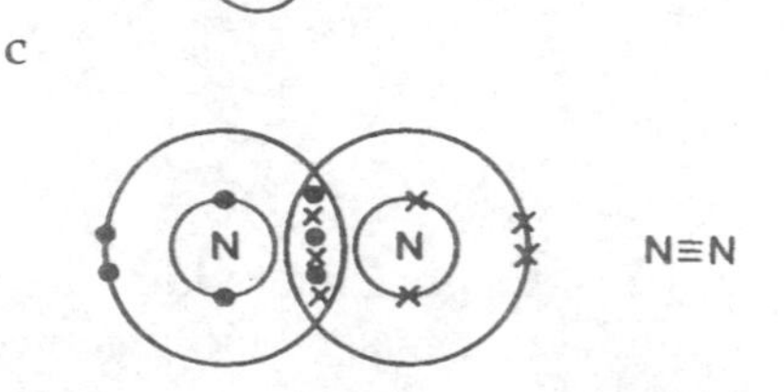

11 Mercury, Hg, a liquid!

12 A metallic; B molecular; C ionic; D giant covalent

Chapter 3

1 a One atom of potassium to one atom of fluorine
 b One atom of sodium to one atom of chlorine
 c Two atoms of carbon to six atoms of hydrogen
 d Six atoms of carbon to five atoms of hydrogen to one atom of nitrogen to two atoms of oxygen.

2 a Two atoms of nitrogen to eight atoms of hydrogen to one atom of sulphur to four atoms of oxygen
 b One atom of magnesium to two atoms of nitrogen to six atoms of oxygen
 c Three atoms of calcium to two atoms of phosphorus to eight atoms of oxygen

3 a lithium bromide; b sodium oxide;
 c magnesium oxide; d magnesium iodide;
 e iron sulphide; f aluminium oxide

4 a lithium carbonate; b calcium hydroxide;
 c aluminium phosphate; d magnesium sulphate;
 e copper nitrate; f ammonium chloride

5 a Li_2CO_3; b AlI_3; c $NaNO_3$; d $ZnSO_4$; e $(NH_4)_3PO_4$;
 f CaO; g $CuCl_2$; h $AlPO_4$

6 a 2,5; three; b 2,4; four; c 2,8,7; one; d 2,8,6; two;
 e 2,8,8; zero; f 2,8,4; four

7 a

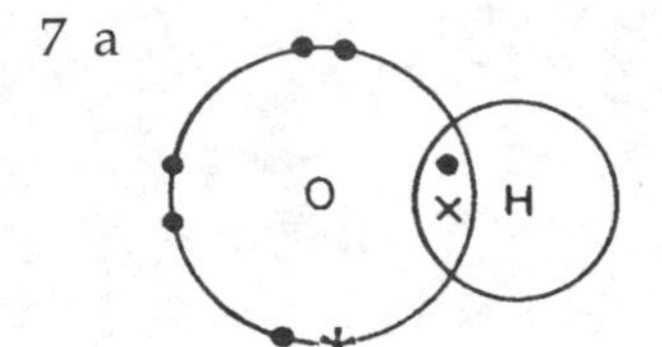

b

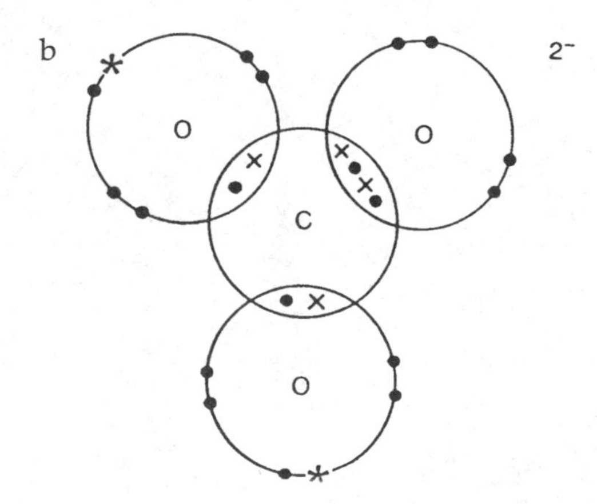

Chapter 4

1 a magnesium + oxygen → magnesium oxide
 b (s) is solid, (l) is liquid, (g) is gas and (aq) is aqueous which means dissolved in water

2 a and d

3 a $2Li + F_2 \rightarrow 2LiF$
 b $4Fe + 3O_2 \rightarrow 2Fe_2O_3$

4 a $Mg + 2HCl \rightarrow MgCl_2 + H_2$
 b $4Na + O_2 \rightarrow 2Na_2O$
 c $Ca(OH)_2 + 2HNO_3 \rightarrow Ca(NO_3)_2 + 2H_2O$
 d $Ca + 2H_2O \rightarrow Ca(OH)_2 + H_2$

5 a iron + sulphur → iron sulphide
 $Fe + S \rightarrow FeS$
 b magnesium + nitrogen → magnesium nitride
 $3Mg + N_2 \rightarrow Mg_3N_2$
 c sodium + bromine → sodium bromide
 $2Na + Br_2 \rightarrow 2NaBr$

6 a $2Mg + O_2 \rightarrow 2MgO$
 b $4Li + O_2 \rightarrow 2Li_2O$
 c $4Al + 3O_2 \rightarrow 2Al_2O_3$
 d $C + O_2 \rightarrow CO_2$
 e $2H_2 + O_2 \rightarrow 2H_2O$

7 $Na \rightarrow Na^+$, so it is oxidised (loss of an electron);
 $Cl \rightarrow Cl^-$, so it is reduced (gain of an electron).

8 a $Mg + 2HCl \rightarrow MgCl_2 + H_2$
 $Mg + 2HNO_3 \rightarrow Mg(NO_3)_2 + H_2$
 $Mg + H_2SO_4 \rightarrow MgSO_4 + H_2$
 $Mg + 2CH_3COOH \rightarrow Mg(CH_3COO)_2 + H_2$
 b $NaOH + HCl \rightarrow NaCl + H_2O$
 $NaOH + HNO_3 \rightarrow NaNO_3 + H_2O$
 $2NaOH + H_2SO_4 \rightarrow Na_2SO_4 + 2H_2O$
 $NaOH + CH_3COOH \rightarrow NaCH_3COO + H_2O$
 c $Na_2CO_3 + 2HCl \rightarrow 2NaCl + CO_2 + H_2O$
 $Na_2CO_3 + 2HNO_3 \rightarrow 2NaNO_3 + CO_2 + H_2O$
 $Na_2CO_3 + H_2SO_4 \rightarrow Na_2SO_4 + CO_2 + H_2O$
 $Na_2CO_3 + 2CH_3COOH \rightarrow 2NaCH_3COO + CO_2 + H_2O$
 $MgCO_3 + 2HCl \rightarrow MgCl_2 + CO_2 + H_2O$
 $MgCO_3 + 2HNO_3 \rightarrow Mg(NO_3)_2 + CO_2 + H_2O$
 $MgCO_3 + H_2SO_4 \rightarrow MgSO_4 + CO_2 + H_2O$
 $MgCO_3 + 2CH_3COOH \rightarrow Mg(CH_3COO)_2 + CO_2 + H_2O$

9 A (blue) solution is added to a colourless solution and a (green) solid forms.

10 a $Pb(NO_3)_2(aq) + 2KI(aq) \rightarrow 2KNO_3(aq) + PbI_2(s)$
 b $K_2CO_3(aq) + BaCl_2(aq) \rightarrow 2KCl(aq) + BaCO_3(s)$
 c $NaCl(aq) + AgNO_3(aq) \rightarrow NaNO_3(aq) + AgCl(s)$

11 Copper is not reactive enough to displace magnesium from its compounds.

12 a Increasing temperature decreases the percentage of ammonia.
 b Increasing pressure increases the percentage of ammonia.

Chapter 5

1 Metals: Sr and Tl Non-metals: F, Xe Semi-metal: Ge

2 Na_2O Sodium oxide (look at the diagrams in Chapter 2 if you had difficulty).

$4\,Na(s) + O_2(g) \rightarrow 2\,Na_2O(s)$
If you had problems, the chapter on Equations will help.

3 a sodium + water → sodium hydroxide + hydrogen
$2\,Na(s) + 2\,H_2O(l) \rightarrow 2\,NaOH(aq) + H_2(g)$
b potassium + water → potassium hydroxide + hydrogen
$2\,K(s) + 2\,H_2O(l) \rightarrow 2\,KOH(aq) + H_2(g)$

4 The outermost electron in caesium is even further away from the pull of the nucleus than that in potassium, so caesium is the more reactive.

5 Very roughly: density 2.3 g/cm^3, melting point 17 °C, boiling point 650 °C.

6 The reactivity will increase as we go from beryllium to radium as the outer electrons get further from the nucleus and are lost more easily.

7 Probably a black solid. Very roughly: melting temperature 300 °C, boiling temperature 340 °C.

8 Less reactive as they must gain two electrons rather than one for the halogen.

9 Hydrogen has one electron in its outer shell so it could react by losing one electron like Group I elements or by gaining one like Group VII elements to get a full outer shell.

Chapter 6

1 a 23; b 16; c 207; d 35.5

2 Any combination of atoms whose A_rs add up to 24, for example 24 hydrogens, 1 carbon plus 12 hydrogens etc.

3 a 17; b 74; c 32; d 46

4 a $4\,g\,H_2 + 32\,g\,O_2 \rightarrow 36\,g\,H_2O$
b $2\,g\,H_2 + 71\,g\,Cl_2 \rightarrow 73\,g\,HCl$

5 a 119 g; b 17 g; c 44 g; d 58.5 g

6 a 1; b 10; c 0.5; d 0.1

7 a

$CuCO_3 \rightarrow$	CuO	$+ CO_2$	
124 g	80 g	44 g	
1 mole	1 mole	1 mole	

b

Mg	$+ 2\,HCl$	$\rightarrow MgCl_2$	$+ H_2$
24 g	73 g	95 g	2 g
1 mole	2 moles	1 mole	1 mole

c

HCl	$+ NaOH$	$\rightarrow NaCl$	$+ H_2O$
36.5 g	40 g	58.5 g	18 g
1 mole	1 mole	1 mole	1 mole

8 a H_2SO_4; b Na_2O; c NH_3; d HNO_3

9 a 2 mol/l; b 1 mol/l; c 5 mol/l

10 a Dissolve 16 g of copper sulphate in 1 l of solution (or any other quantities in the same proportions).
b 16 g of copper sulphate dissolved in 100 cm^3 of solution.

11 a 0.01; b 0.1; 2×10^{-1}

12 a Dissolve 117 g of sodium chloride in 1 l of solution.
b Dissolve 3.4 g of silver nitrate in 2 l of solution.
c Dissolve 19.75 g of sodium thiosulphate in 250 cm^3 of solution.
d Dissolve 25.4 g of iodine in 100 cm^3 of solution.

13 a 2 moles of potassium hydroxide to 1 of sulphuric acid.
b 1 to 1.

14 a 10; b 5×10^{-3}; c 2×10^{-3}; d None

15 a 192.3 cm^3; b 1000 cm^3; c 0.0025 m^3; d 205.4 cm^3; e 9 cm^3; f 115 cm^3

16 0.2 mol/l

17 0.15 mol/l

18 a 20 cm^3; b 500 cm^3

19 a 0.06 g; b 0.163 g; c 0.20 g; d 0.345 g

20 a 240 cm^3; b 480 cm^3

Chapter 7

1 a 1; b 3; c 2; d 4

2 a CH_3
b CH_4

3 a $C_3H_8 + 5\,O_2 \rightarrow 3\,CO_2 + 4\,H_2O$
b $C_4H_{10} + 6\frac{1}{2}\,O_2 \rightarrow 4\,CO_2 + 5\,H_2O$
c $C_{10}H_{22} + 15\frac{1}{2}\,O_2 \rightarrow 10\,CO_2 + 11\,H_2O$

4

```
    H                      H H H
    |                  H   | | |
  H-C-H      and        >C=C-C-C-H
    |                  H     | |
    H                        H H

    H                   H H H H
    |                   | | | |
  H-C-H      and      H-C-C=C-C-H
    |                   |     |
    H                   H     H

  H     H               H H H
   \   /                | | |
    C=C      and      H-C-C-C-H
   /   \                | | |
  H     H               H H H

    H H                    H H
    | |                H   | |
  H-C-C-H    and        >C=C-C-H
    | |                H     |
    H H                      H
```

5

```
  H     H                         H O-H
   \   /                          | |
    C=C      +  H2O  ------>    H-C-C-H
   /   \                          | |
  H     H                         H H
```

6 a dehydration or elimination
  b addition
  c oxidation
  d elimination
  e substitution

7 a propanoic acid

$$CH_3CH_2CH_2OH \xrightarrow{[+O,\ -2H]} CH_3CH_2COOH$$

  b propene

$$CH_3CH_2CH_2OH \longrightarrow CH_3CH{=}CH_2 + H_2O$$

  c 1,2-dibromopropane

$$CH_3CH{=}CH_2 + Br_2 \longrightarrow CH_3CHBrCH_2Br$$

  d propan-1-ol or propan-2-ol

$$CH_3CH{=}CH_2 + H_2O \longrightarrow CH_3CH(OH)CH_3$$

or $CH_3CH_2CH_2OH$

  e 1-chloropropane

$$CH_3CH_2CH_2CH_2OH + HCl \longrightarrow CH_3CH_2CH_2Cl + H_2O$$

  f none (carbon dioxide and water only)

$$CH_3CH_2CH_2CH_2OH + 6O_2 \longrightarrow 4CO_2 + 5H_2O$$

Chapter 8

1 a

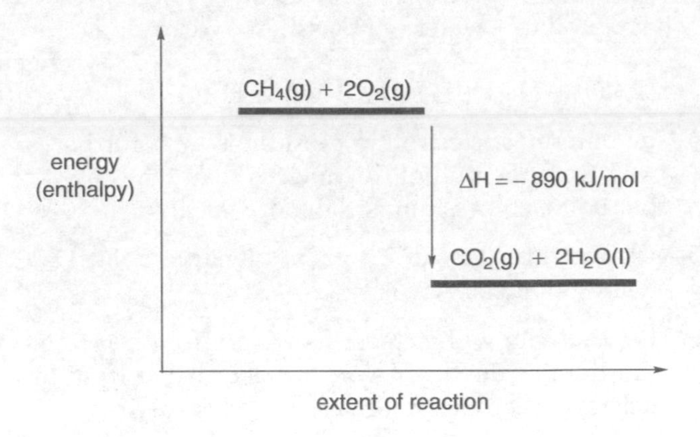

b

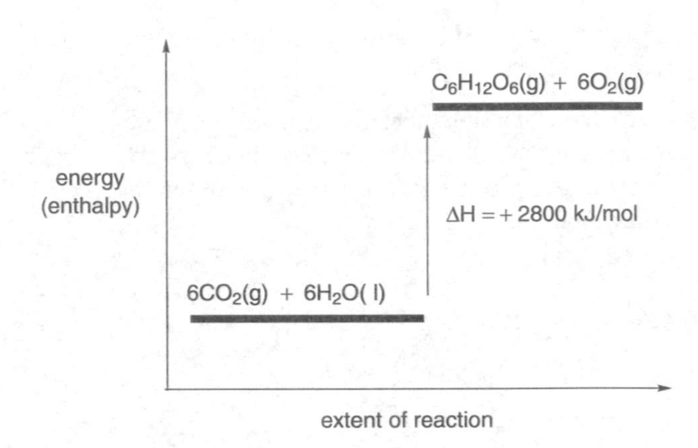

2 – 336 kJ/mol

3 – 546 kJ/mol

4 The thermos flask reduces heat loss by conduction, convection and radiation; the insulation further reduces heat loss by conduction; the narrow neck further reduces heat loss by convection and evaporation.

5 The oxygen supply ensures complete combustion; the copper spiral ensures that heat is transferred to the water.

6 – 1300 kJ/mol

7 – 504 kJ/mol

8 – 2740 kJ/mol

9 – 1724 kJ/mol

10 – 1056 kJ/mol

11 – 122 kJ;/mol (as expected)

12 a – 20 kJ/mol; b – 184 kJ/mol

13 Approximately – 3700 kJ/mol by extrapolation from the graph

14 a C–Cl; b C–H; c O–O

15

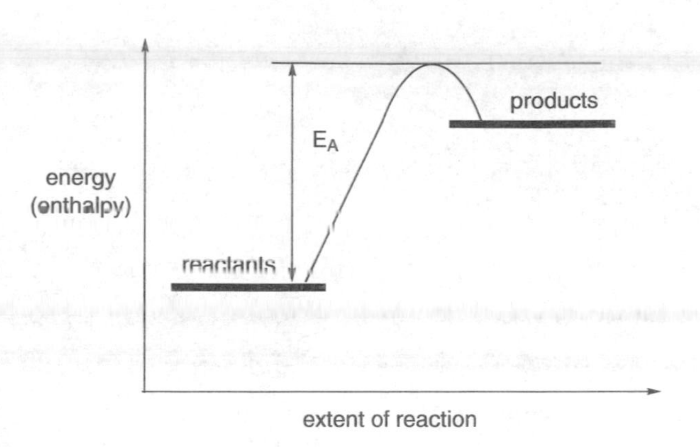

16 a heat from friction; b the spark plug; c sunlight

Chapter 9

1 Metals are good conductors of electricity because they have a 'sea' of electrons which are free to move and carry the current.

2 Ionic compounds contain charged particles whereas covalent compounds do not.

3 Br^-, Ca^{2+}, K^+, F^-, S^{2-}, N^{3-}

4 Calcium and oxygen
a $Ca^{2+} + 2e^- \rightarrow Ca$
$O^{2-} \rightarrow O + 2e^-$
or
$O^{2-} \rightarrow \frac{1}{2}O_2 + 2e^-$
b Aluminium and chlorine
$Al^{3+} + 3e^- \rightarrow Al$
$3Cl^- \rightarrow 1\frac{1}{2}Cl_2 + 3e^-$
c Lead and bromine
$Pb^{2+} + 2e^- \rightarrow Pb$
$2Br^- \rightarrow Br_2 + 2e^-$

5 $H^+ + e^- \rightarrow \frac{1}{2}H_2$
or
$2H^+ + 2e^- \rightarrow H_2$

6 a Hydrogen at the cathode, oxygen at the anode.
b Hydrogen at the cathode, oxygen at the anode.
c Hydrogen at the cathode, oxygen at the anode.

7

Metal	Ion	Half equation at cathode	No. of faradays needed for 1 mole	1 faraday will produce...	
				No. of moles	Mass/g
Ca	Ca^{2+}	$Ca^{2+} + 2e^- \rightarrow Ca$	2	$\frac{1}{2}$	20
Na	Na^+	$Na^+ + e^- \rightarrow Na$	1	1	23
Mg	Mg^{2+}	$Mg^{2+} + 2e^- \rightarrow Mg$	2	$\frac{1}{2}$	12
Al	Al^{3+}	$Al^{3+} + 3e^- \rightarrow Al$	3	$\frac{1}{3}$	9
Cu	Cu^{2+}	$Cu^{2+} + 2e^- \rightarrow Cu$	2	$\frac{1}{2}$	32

8 2 faradays produce 1 mole of calcium as calcium ions are Ca^{2+}; $Ca^{2+} + 2e^- \rightarrow Ca$. 1 mole of calcium has a mass of 40 g.
2 faradays produce 1 mole of hydrogen, H_2, as hydrogen ions are H^+; $2H^+ + 2e^- \rightarrow H_2$. One mole of hydrogen has a mass of 2 g.

9 $Na^+ + e^- \rightarrow Na$
No. of coulombs $= 15\,000 \times 12 \times 60 \times 60$
$= 648\,000\,000$ C
No. of faradays $= 648\,000\,000/96\,000 = 6750$ F
Mass of sodium $= 6750 \times 23 = 155\,250$ g

10 $Cu^{2+} + 2e^- \rightarrow Cu$
No. of coulombs $= 0.1 \times 10 \times 60 = 60$ C
No. of faradays $= 60/96\,000 = 0.000\,625$ F
Mass of copper $= \frac{0.000\,625}{2} \times 64 = 0.01$ g

11 The current (and therefore the no. of faradays) is the same through each part of the apparatus.
No. of coulombs = 576 C
No. of faradays = 0.006 F
For copper:
No. of moles deposited = 0.003 mol
The charge is Cu^{2+}
For silver:
No. of moles deposited = 0.006 mol
The charge is Ag^+

Chapter 10

1 a 2 amps; b 5 amps; c 6 amps

2 a 10 C; b 14 400 C

3 3 J

4 a 0.5 amp; b 0.5 amp; c 0.5 amp

5 30 Ω

6 16 V

7 2 m/s

8 3.3 m

9 In some cases, the waves add together and in others they cancel out.

Chapter 11

1 a It will drop to one third of its original volume.
b It will quadruple.
c $x \propto y$

2 a $P=RT/V$; b $V=RT/P$; c $T=PV/R$;
d $R=PV/T$

3 a $q=pr$; b $m=n/t$; c $h=fe/g$; d $r=soq/p$

4 a 0.2428; b 4.2×10^{-4}; c −6.909; d 0.169

5 a 64; b 256; c 27

6 a 10 000 000 or 10 × 10 × 10 × 10 × 10 × 10 × 10
b 1×10^5 or 10 × 10 × 10 × 10 × 10
c 300 000 or 3×10^5
d 1/10 000 000 or
1/(10 × 10 × 10 × 10 × 10 × 10 × 10)
e 1×10^{-9} or
1/(10 × 10 × 10 × 10 × 10 × 10 × 10 × 10 × 10)
f 10^{-6} or 1/1 000 000

7 a 1×10^8; b 10; c 1×10^{-2}

8 a 2.33×10^3; b 3×10^{12}; c 4.34×10^{46}

9 6

10 pH=4

11 a 1.146; b 0.845; c 0.995; d −0.155; e −2.523

12 a 50.118; b 630.957; c 7943.282; d 1.58; e 1.175;
f 1.022

13 a 280 g; b 32 g

14 3 cm^3/s

15 Approximately 0.33 cm^3/s